A Bibliography of
FIREWORK BOOKS

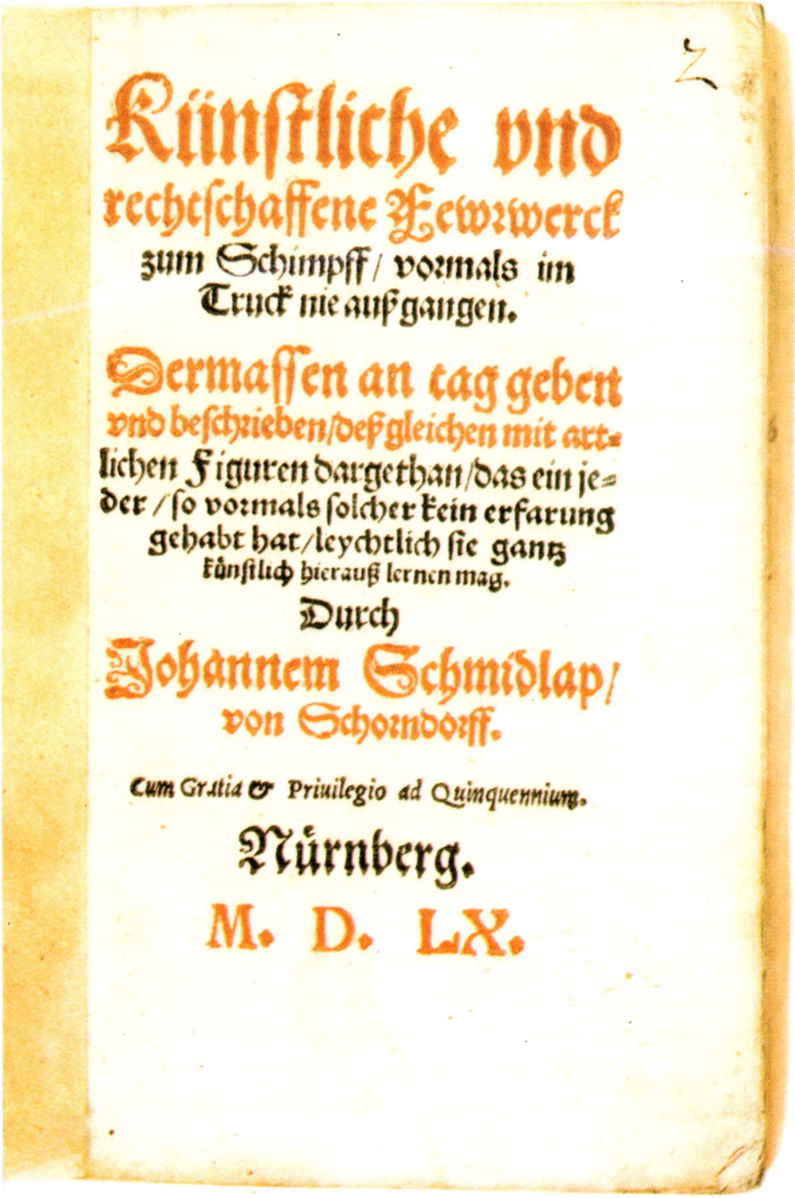
Künstliche vnd
rechtschaffene Fewrwerck
zum Schimpff/ vormals im
Truck nie außgangen.

Dermassen an tag geben
vnd beschrieben/deßgleichen mit art-
lichen Figuren dargethan/das ein je-
der/so vormals solcher kein erfarung
gehabt hat/leychtlich sie gantz
künstlich hierauß lernen mag.

Durch
Johannem Schmidlap/
von Schorndorff.

Cum Gratia & Priuilegio ad Quinquennium.

Nürnberg.

M. D. LX.

Title-page of the first edition of the first book to deal exclusively with recreative fireworks.
(Courtesy of the Royal Danish Military Library, Copenhagen)

A Bibliography of

FIREWORK BOOKS

Works on recreative fireworks

from the sixteenth to the twentieth century

by Chris Philip

Published by C. Philip,

in association with St. Paul's Bibliographies

1985

British Library Cataloguing in Publication Data

Philip C.
A bibliography of firework books: works on recreative fireworks from the sixteenth to the twentieth century.
1. Fireworks——Bibliography
I. Title
061.662'1 Z5885

ISBN 0-906795-31-1

Published by C. Philip, in association with
St. Paul's Bibliographies, West End House, 1 Step Terrace, Winchester, Hampshire SO22 5BW

Technical note

The text was compiled on a Tandy TRS-80 Model III using SCRIPSIT enhanced by POWERSCRIPT to enable all the International characters and facsimile type faces to be generated on an EPSON FX-80. The floppy discs were then transferred to and made compatible with an IBM PC by using XENO-COPY. Typesetting was done on a LINOTRON 202, using CORA interface software, by Serious Software of London.

Printed in Great Britain by Henry Ling Ltd., The Dorset Press, Dorchester.

To the Rev. Ronald Lancaster

for his help and encouragement

Contents

Author's note

I am of course well aware of the various omissions and shortcomings that inevitably accompany the first edition of a work of this kind. I would therefore ask that any reader who knows of any, or can assist in any way, write to me at the Publisher's address and, if possible, send a photocopy of the relevant titlepage. All such contributions will be incorporated in any future edition and be gratefully acknowledged.

Abbreviations

ads.	Advertisement
amp.	Ampliata. (Enlarged)
aufl.	Auflage. (Edition)
aug.	Augmente. (Enlarged)
ausg.	Ausgabe. (Edition)
bear.	Bearbeiten. (Revised)
BL...	British Library (followed by shelf-mark if held)
BR	Black and red printed title-page
EngTP	Engraved title-page
F	Folding
figs.	Figures
illus.	Illustrated
inc.	Including
IOL	India Office Library
L	Leaves
N...	National Union Catalogue reference number. (See 'BIBLIOGRAPHIC NOTES')
nd.	No date
np.	No printer
p.	Page(s)
PC	Paper covers
Pls.	Plates
PO	Patent Office Library. (Now Science Reference Library, London)
riv.	Riveduta (Revised)
Tab.	Tables
TP	Title-page
umgear.	Umgearbeitet (Revised)
verb.	Verbessen. (Improved)
verm.	Vermehrte. (Enlarged)
WC	Woodcuts

List of illustrations

All the illustrations are from the author's own collection, except for the frontispiece, which was kindly provided by the Royal Danish Military Library, Copenhagen.

The illustrations are arranged in chronological order of the first edition in which they appeared. In some instances the illustrations in this bibliography have been taken from later editions; in these cases the date of the actual edition used is given in brackets. For all practical purposes however they may be considered to be identical to those printed in the first edition.

Preface

'It seems to me that the justification for any bibliography is that the information contained in it has not hitherto been readily available to the student and the collector.' Trevor H. Hall, *A Bibliography of Books on Conjuring in English from 1580 to 1850*. Minneapolis, 1957.

I first became fascinated by fireworks in the late 1960s when I used to visit Malta and her sister island Gozo, where I had acquired an old farmhouse. On nearly every summer weekend, one or other of the small villages would have a firework display to celebrate their patronal feast day, each village vying with the other to produce louder and better shells than the next.

As the quality of their fireworks was so much better than those that I had hitherto seen in England and, because I imagined that they must be comparatively cheap since Malta is rather a poor island, I was determined to start importing them into England. Luckily, my ignorance at that time of everything pyrotechnic was to stand me in good stead, because I was unaware of all the difficulties that I would have to overcome. Had I known then what I know now, I might not have tried!

In order to learn more about fireworks I started looking for books on the subject. It was at about this time that I first met the Rev. Ron Lancaster and he gave me a copy of his book that had just been published entitled *Fireworks: Principles and Practice*. He also suggested that I should obtain a copy of Brock's *History of Fireworks*. With these two books as starters, I soon became fascinated by the subject and began to look for other books on the subject.

The first item I came across was really a 'Fête' book, called *Relation de l'Inauguration solomnelle de sa Sacre Majestie Imperiale et Catholique Charles VI Empereur des Romains . . . comme Comte de Flanders. Celebré à GAND Octobre 1717*. The four spectacular firework plates it contained completely captivated me.

The next item I found was Alberti's *La Pirotechnia* but the shop wanted £250; a great deal of money I thought for 1972. By the time I saved up enough, it had been sold – but the frustration spurred me to become a collector.

Apart from the rather brief and incomplete bibliography at the back of Brock's book, I was unable to find any other reliable work on the subject of recreative firework books and thus I started to compile this Bibliography. Never having attempted such a thing before and having no one to guide me I started at what was, to me, the most obvious point, which was Peddie's *Subject Guide;* having noted down all the entries under "Fireworks" I then obtained a reader's ticket for the British Library and examined all the books he listed. At the same time I examined all the other editions of the same author as well as noting all the entries given in the National Union Catalog.

At this time I had very little idea what it was I should be making notes about; which items were of significance and which were not. Gradually, however, some pattern and shape emerged and the present format and method of description was developed. I should perhaps stress at this point that the layout and description used throughout this work does not always conform to standard bibliographic practice, so I would urge readers to study the notes carefully.

Many unexpected titles were found by the somewhat haphazard method of taking likely-looking reference books off the shelves of the British Library and looking up "Fireworks" or "Pyrotechnics" in the index. A rather more methodical approach was adopted after I discovered a *Guide to Reference Books* (American Library Association: Chicago 1976). This book at least told me the most likely sources for the material I was seeking.

I make no claim to having examined personally all the titles listed. An attempt to examine those which are available at libraries other than the British Library would have taken a lifetime, apart from the cost involved. However, as I possess approximately a third of the over five hundred entries in the present work, which represents about 90% of all the authors involved, and as I have examined most of the ones held by the British Library, I think I can claim to have covered the greater part of the whole. In many cases it would seem that none of the major libraries in America or Britain possesses some of the editions quoted, although of course they might be available in one or other of the European libraries. Furthermore, I suspect that a lengthy and costly search for an elusive edition would be unrewarding. With the numerous editions published by some of the prolific but minor mid-nineteenth century German authors, such as Loden, Weber or Buettner/Sandmann, I think it would be a case of if you've seen one, you've seen them all.

There are, despite my efforts, a large number of blanks left in the text. These have been left there deliberately in order to encourage people to help me complete information that I have been unable to find. During my work on the bibliography, I was told a salutary tale by one of the London booksellers whom I regularly frequent. One of his friends had spent a lifetime working on a bibliography, but had never been persuaded to publish it because he always felt that there was still some small item to be included, or some small query to be resolved. As a result he was now both too old and too impoverished to undertake the publication of his life's work. I have no intention of allowing this work to suffer a similar fate!

I hope, therefore, that those who criticise this work will also help me to fill the inevitable gaps in order that a more complete second edition can be prepared. I know that my pyrotechnic readers will probably complain that I have not given sufficient emphasis to the scientific and historical side of the subject; but if, also, the bibliographer complains that I have neglected certain aspects of his own discipline then I will probably have achieved a reasonable balance.

Another reason for undertaking this work is the growing scarcity of the books listed. In the 1920s Sotheran's catalogue offered literally dozens of books on fireworks and, even in the 1960s and 1970s one could be offered several books on the subject. However, during recent years, the number of books being offered has declined sharply and, despite being in regular contact with a large number of bookdealers both in England and Europe, I find that fewer and fewer firework books become available for sale. Indeed many booksellers that I approached with the query: "Do you have any books on fireworks?", often expressed the opinion that with the possible exception of Brock's *A History of Fireworks*, there probably weren't any and if there were, they had never possessed one in their lifetime! In some ten years of collecting, virtually the only firework books that have been offered to me more than once are, apart from Brock's, those by Frezier, Browne, Capt. Jones and Kentish. Even so, I am still without the elusive first edition of Frezier published in 1706. I think that this scarcity is clearly demonstrated by a glance at the large number of titles and editions that are *not* represented in either the National Union Catalog or the British Library Catalogue. It is in order to highlight this that I have deliberately left blanks against these two catalogue references in the main text.

Whilst I was undertaking this research I read a number of works dealing with bibliography and in particular took to heart the questions posed by Philip Gaskell in his *A New Introduction to Bibliography* (O.U.P. 1972) in which he says: 'In deciding what and how much to include, the bibliographer must ask himself repeatedly: "What is the purpose of the description? Who really needs each item of information? Can anything be abbreviated?" Only thus can he avoid burdensome and expensive superfluity, and escape the ultimate absurdity of mistaking the means of bibliography for its end, of practising bibliography for bibliography's sake.'

With these questions in mind therefore, I offer this work as a 'list' of books – I hesitate to call it a 'bibliography' – that deal with recreative fireworks, for the benefit of collectors, book-sellers, reference libraries but, above all, for people who, like myself, love fireworks.

Acknowledgements

Book-collectors need book-sellers to survive. So, first and foremost, I must acknowledge my indebtedness to the fifty or so book-sellers from around the world who have helped me over the last dozen or so years to build up my collection of firework books. Without this foundation, this bibliography could not have been compiled. I cannot, of course, list them all, but special mention must be made of Paul Breman, Michael Phelps, Graham Weiner, Bill & Keith Fletcher and Roger Gaskell; all of whom have gone out of their way to impart knowledge and advice, as well as to sell me their books.

The idea of writing the bibliography itself primarily arose from the many years of correspondence with Bob Cardwell, the publisher of *Pyrotechnica* in Austin, Texas, USA, who has devoted much time to correcting proofs and supplying title information. I am also indebted to *Pyrotechnica* (Vol.IX) for Alex Schumann's translation of Takeo Shimizu's notes on Japanese firework books. Also from America, Ray Anderson, who kindly supplied me with photocopies of the catalogue cards of the Oscar Guttmann collection in the Du Pont Eleutherian Mills Library, Delaware. Particular thanks are due to Michael Swisher of Stillwater, Minnesota, USA, who very kindly read through the final proofs and provided a great deal of information about some of the leading pyrotechnic writers of the past and whose comments on Frezier, Perrinet d'Orval, Chertier, Tessier and others I have freely quoted. In addition he also supplied me with a transliteration of Shidlovsky's list of early Russian books on pyrotechnics.

I am also most grateful to Professor John Horden, Head of the Bibliographic Department of Stirling University, Scotland, for his enthusiastic support of this venture.

Having boldly decided to compile and typeset the entire work on a computer, partly to save money and partly from the enjoyment and interest it gave me, I must also thank John Newgas, of the local Tandy User Group, who helped solve a number of intractable problems concerned with arcane matters such as interfacing, bytes, ASCII and Hexadecimal numbers, and also Derek Kay of Serious Software who was able to solve the fundamental problem of transferring Tandy TRSDOS data direct to their typesetting machine.

A number of firework companies in Europe have collections of books on the subject that their families have built up over the centuries and one of the most extensive that I have had the pleasure of examining is that of

Lars Barfod of Copenhagen, who is responsible for the firework displays at the Tivoli Gardens. My visit to Copenhagen was memorable in more ways than one, since he introduced me not only to The Royal Danish Library but also to The Royal Danish Military Library amongst whose extensive and superbly kept collection of early military books and manuscripts I discovered the only known extant copy of Schmidlap's *Künstliche vnd rechtschaffene Fewrwerk zum Schimpff* (1560); the first book published anywhere dealing only with fireworks for pleasure, as opposed to their military use. A fitting climax to my work on this book and I am most grateful to Lt. Col. Christensen and his staff for their help and kindness.

I must not of course forget the enquiry staff of the British Library who have answered all my queries over the last few years with patience and forebearance. Also the staff of The Royal Artillery Institution Library at Woolwich for allowing me to examine their private collection.

Finally I must acknowledge my debt to the staff of St. Paul's Bibliographies who have guided my footsteps along the paths of publishing.

Introduction

This bibliography of books on fireworks is concerned primarily with 'recreative' rather than 'military' fireworks, 'pour la recréation' or 'pour la joie' rather than 'pour la guerre', for 'die Lustfeuerwerker' and not 'der Ernstfeuerwerker'. I must apologise for the use of this rather cumbersome word 'recreative', but I have not been able to think of a more suitable one. The literal translation of 'die Lustfeuerwerk' or 'feux d'artifice pour la joie' as 'fireworks for joy or pleasure' does not seem altogether suitable and as the intent of the work is to differentiate clearly between war-like uses and peaceful uses, 'fireworks' by itself is not sufficient. It is, however, virtually impossible to separate the military from the peaceful use of fireworks prior to about the mid-17th century, so quite a number of books on artillery are included. In some cases, because the long-title alone is not sufficient to give a clue as to whether or not a military book contains recreative fireworks a number of such works have been included and the lack or presence of 'fireworks for triumph or peace' noted. This field, and the associated one of Greek fire and gunpowder, are already well covered by writers such as Cockle's *Bibliography of Military Books up to 1642*, Jähns's *Geschichte der Kriegswissenscften*, and Partington's *A History of Greek Fire and Gunpowder*. Guttmann's *The Manufacture of Explosives* (1895) has a substantial bibliography on the subject. Some overlapping is therefore inevitable.

I should perhaps list the items that this bibliography does not contain. It does not list general encyclopaedias, technical encyclopaedias or technical journals which contain articles on the chemistry of pyrotechnics, nor does it list magazines that have isolated articles on fireworks. It does, however, include a number of magazines devoted exclusively to the subject of pyrotechnics and these are all to be found under MAGAZINES. It is perhaps strange, that, with the exception of a series of German magazines published in Weimar in the 1840s and 1850s, there were no others in any language until the American Pyronews commenced publication in 1966. Where authors have written books that deal collectively with military and with recreative fireworks, both have generally been included but their other works have not. (You will not find Captain Jones's books on skating listed!) I have not listed ephemeral items such as manufacturer's and exhibition catalogues, Fête books, or programmes and leaflets describing specific displays. There is plenty of work left for others who are interested in these aspects of the subject.

In compiling this bibliography, it soon became apparent that nearly all books dealing with the subject of recreative fireworks fell into four clearly defined types. Firstly, there were those that dealt solely and exclusively with the subject of fireworks for pleasurable purposes. Secondly, there were those which dealt with artillery but had a section at the end dealing with recreative fireworks. The third group is what I call 'Mathematical Recreations' because the first of them to include a section on firework was van Etten's *Récréations Mathematiques* (1628). This genre of book appears in dozens of guises during the next three hundred years. The last group is harder to define as it includes not only books which deal with other related subjects such as conjuring but also chap-books. I will discuss each of these groups in greater detail.

Recreative Fireworks

The first book to mention the subject is probably Biringuccio's *Pirotecnia*(1540). Although the greater part of this work deals with Metallurgy, 'Book X' deals with Gunpowder, Greek-fire, Fire-tubes and the like, and it also mentions the use of Fireworks on 'solomn feast days' in Tuscany. Valle's Libro Continente (1524) also describes Fire-tubes and Wildfire but only for war. During the latter half of the 16th century and the early part of the 17th, the many books dealing primarily with artillery devote more and more pages to the peaceful uses of the Art. The first book to deal exclusively with the use of fireworks for recreational purposes was probably that of Johann Schmidlap. His *Künstliche vnd rechtschaffene Fewerweck* appeared in 1560, published in Nuremberg. However one of the most influential was Ufano's *Tratado de la Artilleria* (1612), which provided inspiration, both acknowledged and plagiarised, for numerous later works. Some even acknowledged the fact. Its plates, which were engraved by de Bry, were still being used by others sixty years later. It is not however until we get to Babington's *Pyrotechnia* of 1635 that the first book in English to deal solely with the subject of recreative fireworks appears. In many ways it stands head and shoulders above the rest. It was to occupy a unique place in pyrotechnic literature for another 71 years, until the first major French work on the subject, Frezier's *Traité des Feux d'Artifice* (1706). It treats its subject in great detail and with great imagination and was a source of inspiration for many subsequent authors.

In 1749 the first Italian work devoted solely to the subject was published, Alberti's *La Pirotechnia*. It was not however until 130 years after the publication of Babington's work that the second English book devoted exclusively to recreative fireworks appeared, in 1765, written by Captain Jones, entitled *A new treatise on Artificial Fireworks.* America had to wait until Cutbush's *A History of Pyrotechny* was published in 1825. The 19th century saw the appearance of large numbers of firework books. For many, it became an intriguing hobby, until legislation put a stop to amateur manufacture in many parts of the world. The 20th century landmarks, apart from Brock's book which deals, like no other had up to then, with a broad-sweeping history of the subject, are Weingart's *Dictionary of Pyrotechnics* of 1930, Lancaster's *Fireworks. Principles and Practice* of 1972, and the various works of Shimizu whose exceptionally detailed research is unsurpassed.

Artillery and Military

As already mentioned, because recreative fireworks developed from the war-like uses of gunpowder and fire, many of the early books on warfare and artillery contained sections devoted to the peaceful uses of fireworks. Even in our own century, there are many titles which deal with pyrotechnics for both war and peace, such as those by Ellern and Davis. In addition there are also a number of compendium works which included a bit of everything that the skilled soldier or sailor was expected to be able to turn his hand to. These range from Sturmy's *Mariner's Magazine* (1669), and Venn's *Military and Maritine (sic) Discipline* (1672) to Capt. Griffith's *The Artillerist's Manual* of 1839, which went through ten editions up to 1868 and included a section on fireworks, because 'from his own personal knowledge ... Firemasters and Laboratory Serjeants have been called on to Superintend the construction, and exhibit a Display of Artificial Fireworks.'

Mathematical Recreations

In 1628 Charles Osmont published in Rouen the first of what was to prove, in many ways, the most popular type of book containing details of how to make fireworks that was ever printed. This was *Récréations Mathematiques* by Henri van Etten, an enlarged version, in three parts, of *Récréation Mathematique* which had first been published by Jean Appier in Pont-à-Mousson in 1624 and whose popularity was such that at least half-a-dozen other editions and versions had already been published in the four years since it had first appeared.

This new third part, entitled 'un recueil de plusieurs gentilles & recreative inventions de feux d'artifice', was probably contributed by Appier who had also written *Recueil de plusieurs Machines Militaires et feux Artificials* in 1620. *Récréation Mathematique* was a compilation, by van Etten, of what were then considered tricks, scientific curiosities, games and such like which were part truth and part myth but whose main purpose was to entertain the reader.

This type of book was immensely popular and as a result there were dozens of imitations, commentaries and translations. The first of the commentators was signed only D.A.L.G. (1626), followed by D.H.P.E.M.

(Denis Henrion, Professor of Mathematics) also in 1626, and published under the title *Examen du Livre des Récréations Mathematiques*. Several reissues were made until in 1659 when it appeared under the title *Les Récréations Mathematiques .... Premierement revue par D.Henrion, depuis par M.Mydorge.* This, in turn, went through various editions and reissues until 1743. In parallel with these annotated versions the original work, without comments, also went through a large number of editions and reissues until 1680.

In 1633, the first English translation appeared under the title *Mathematicall Recreations*. This translation is erroneously ascribed to Oughtred. It is probably by Malthus, and two further editions of it appeared in 1653 and 1674. The reasoning behind this theory is discussed in great detail by Trevor Hall in his essay *Mathematicall Recreations*. An exercise in *Seventeenth Century Bibliography*, University of Leeds (1969).

The next series of books of a similar nature were those originally compiled by Jacques Ozanam who published a two volume edition called *Récréations Mathematiques et Physiques* in Amsterdam in 1693. This in turn went through over a dozen versions printed both in Amsterdam and Paris until 1750, a four volume version first appearing in 1723. In 1778 a new edition appeared 'considerablement augmentee par M. de C.g.f.', the pseudonym of J.E. Montucla. A further edition appeared in 1790.

The first English translation of Ozanam's version, by an unknown author, was made in 1708 and three subsequent editions, under the title *Recreations for Gentlemen and Ladies* appeared in 1756, 1759 and 1790. It was however much abbreviated and consists of only some 200 pages.

A close translation, by Hutton, of the 1790 Montucla edition of Ozanam, first appeared in 1803 under the title *Mathematics and Natural Philosophy*. In his introduction Hutton makes a very significant comment. 'Why it has been usual to consider pyrotechny as a branch of the mathematics, we do not know ... However, as we might be blamed for omitting an art which affords a considerable field for amusement, and as it is connected at least with natural philosophy, we shall make it the subject of one division of this work'. The section on fireworks having therefore been included in the edition of *Récréations Mathematiques* of 1628, no subsequent author has dared to leave them out, no matter how removed from the main subject of the work they may appear! In addition to these versions, *Récréations Mathematiques* has also been translated both into German, by Schwenters, and into Dutch by Westen. In all these versions, the original section on recreative fireworks appears in some form or another.

Apart from the sequences of editions already mentioned, other authors such as Guyot and Hooper also copied the general format of *Recreation Mathematique* although in their versions the section on fireworks is confined only to imitations of fireworks by using lights behind paper cut-outs and, as such, not strictly relevant to the subject of this Bibliography.

There is however yet another series of books that contains sections on recreative fireworks and that is the conjuring books of John White and his imitators. His *A Rich Cabinet ... for the recreation of ingenious spirits in their vacant houres*, first published in 1651 contains, apart from many sections devoted to 'Hocus-pocus' and 'Leger-de-Main', a substantial section on fireworks. In what sometimes appears to be a desperate bid not, as Hutton would have it, 'to be blamed for omitting an art that affords a considerable field for amusement', White includes sections on 'Fire-pikes' and 'How to invest ships' next to items devoted to killing mice and how to make one's hair grow longer. Similarly, as a result of White's inclusion of fireworks in his *Rich Cabinet*, many subsequent works on conjuring also include short paragraphs on pyrotechnics although the details given would not be of much help to a serious student.

It is perhaps ironic that, despite the great popularity of these books, they are today very scarce. In 1860, the great bibliographer A. Deblaye wrote, 'Tous ces ouvrages sont fort rares, quelques-uns presqu' introuvables'. It is probable that their very popularity literally wore the books out and they were then thrown away, whereas the learned leather-bound philosophical treatise looked good on the bookshelf but was seldom read and has thus survived.

Miscellaneous

A number of titles do not of course fit easily into any of the three preceding categories. Into this class, I would put the innumerable chap-books that appeared in the late 18th and early 19th century. These were small, stitched tracts sold cheaply by itinerant dealers or 'chapmen'; they usually comprised multiples of 4 pages and were illustrated with primitive woodcuts. They were of little literary merit as the texts were crude, anonymous and probably plagiarised. By the end of the 19th century, with the advent of magazines and the expansion of publishing, they had virtually disappeared. Priced at six-pence and consisting of perhaps 12 or 32 pages, they had grandiose titles such as 'The Whole Art of ...' but in reality would prove of little practical use to the serious worker. Also in this category are what might better be called 'peripherals'. These are books that, although mentioned in other works on pyrotechnics, are not strictly relevant to the subject of this bibliography. They have been included for the sake of completeness, but usually have an accompanying note explaining their contents. In some cases, the book's title may have misled previous authors into thinking they might be relevant. Starkey's works are a good example. Others, like Guttmann's, which all deal with explosives in general, nevertheless have strong connections with our subject.

Bibliographic Notes

I would particularly ask readers to study the following notes carefully because some of the bibliographic methods and descriptions used are slightly unorthodox and do not always follow established practice.

Name

The author's name is generally as shown and spelt in the National Union Catalog, together with the forenames when known. In cases where there might be doubt or confusion, appropriate cross-references are also given. When the name of the author is not given in the book or is not known, the listing is under the first main-word of the title, excluding the article. A full list of all such books is given under ANONYMOUS. When the name of the Author is not given but is generally known the title is cross-referenced to the Author.

Date

The author's dates are those given in the NUC. I have not adopted their convention of using 'fl.' for 'flourished'. I am prepared to assume that the author generally did just that, at the time his book was published.

Title

As early books tended to have very long titles - indeed almost an index - abbreviated titles are generally shown except when all or part of the long-title is a helpful guide to contents, or in determining the variations in editions or issues. The original spellings are shown, except for the use of the modern 's'. Many early title-pages suffered from so many curious, erratic and incorrect spellings that I have refrained from what would otherwise be an all too liberal use of 'sic'. Furthermore, in many early French and German books, accents were not used as consistently as they are today. I hope, therefore, that all apparent mistakes in titles are as in the original. As it was obviously necessary to provide English transliterations and translations for the Japanese and Russian titles I thought it would be helpful to provide a rough translation for most of the other foreign titles which might otherwise not be clear to all readers.

Reference Number

The numbering system I have adopted is a mixture of that used by Wing and the STC. It is designed to permit the inclusion at some future date of additional authors, issues or editions without upsetting the sequence and, at the same time, providing some additional information about the volume concerned. Each letter of the alphabet under which the work is listed, either by the initial letter of the author's surname or, if anonymous, by the initial letter of the first word excluding the indefinite article, consists of a self-contained sequence which runs A010, A020, A030, A040 etc. and B010, B020, B030 etc. Each issue or edition under that author is then given a decimal number as A010.1, A010.2, A010.3. Thus a reference number 'E090.15' would indicate that there were fourteen previous issues or editions of that particular author.

Publisher

The name of the publisher is given as shown on the title-page (sometimes slightly abbreviated). Occasionally, this name is also that of the printer and/or seller.

Collation

See list of abbreviations for details. When the main text page numbering does not start at '1', a negative figure in brackets after the numbered pages is shown. For example, 274(-6), indicates that there are six preliminary pages, (frequently numbered, if at all, in Roman numerals), with the text starting on a page numbered 7.

Town

The publisher's town (or sometimes towns) is also given as shown on the title-page. The modern name is given in brackets.

Date

The date is shown as given on the title-page in either Roman or Arabic. If given in Roman, the Arabic is shown in brackets afterwards. When no date is given on the title-page, but a date is shown elsewhere in the book, this is printed in round brackets ( ). Where no date is given anywhere in the book, but it is known from other sources, square brackets [ ] are used. Where no precise information is available, 'c.' (circa) is used.

Size

Rather than using the somewhat vague and imprecise method of denoting size by 8vo. or Folio etc., the height of all copies that have been examined were measured in centimetres, this being in line with the NUC method. It should of course be realised that variations in trimming can easily cause differences between otherwise identical volumes of 2 or 3 centimetres either way.

Catalogue Numbers

The National Union Catalog number is given where applicable. This is shown as 'N' followed by the initial letter of the author's name. This sometimes differs from the name under which the entries are here catalogued, (e.g. NUC lists HUTTON's translations of OZANAM under 'OZANAM', hence the reference is given in the NUC catalogue as 'NO' and not 'NH'). The use of the prefix 'NS', followed by the initial letter of the author, indicates the NUC Supplement which commenced publication in 1980 and, like the NUC itself, deals only with pre-1956 imprints. The use of the prefix 'RL' refers to the Rowan & Littlefield NUC Supplement (1956-1967). The British Library shelf-mark is included as given in the *General Catalogue of Printed Books* in the Main Reading Room. This sometimes differs from the number shown in the published volumes. A '+' indicates that another copy is available. A 'D' indicates the volume was destroyed during the war but the number may still be shown in some published editions of the Catalogue that have not been amended.

A hyphen following the 'N' prefix indicates that the book in question probably does not exist in any major American Library. Similarly a hyphen after 'BL' denotes that it is probably not in the British Library. These have been put in deliberately to indicate the large number of works and editions that do not appear in either of these two great catalogues. I use the word 'probably', because there are differences in cataloguing methods between these systems. For example, for a long time I did not think that the anonymous work *The Perfection of Military Discipline* existed in the British Library. In the NUC it appears under PERFECTION, in the BL it appears under ENGLAND because the long-title is *The Perfection of Military Discipline, after the newest method, as practised in England and Ireland.*

Edition

Where there are a number of differing editions, the title-page comment is shown, sometimes abbreviated, e.g. 'Seconda ed. riv. & aum.'. If no comment is shown on the title-page, the information is interpolated and shown in brackets, thus - (2nd ed.).

Notes

Where there are notes at the end of the Author entry, I have attempted wherever possible to quote from other authors and works who have had something to say about the work in question. Hodgkin's *Rariora* and Sotheran's *Bibliotheca Chemico-Mathimatica* have provided a great deal of useful comment. All references in these Notes to authors who have an entry in this Bibliography are printed in capitals.

References and Locations

Outside the British Library, the Du Pont Eleutherian Mills Library in Willmington, Delaware, USA, probably has the largest collection of firework books available to the public. It consists in the main of the Oscar Guttmann library. Most of the titles from its index cards are included, although a few that deal only with military fireworks are not noted.

The STC numbers of Pollard & Redgrave and Wing have been given where applicable, as well as the dozen or so titles in the 'Berlin' catalogue that is often referred to by booksellers.

At the beginning of 1984 I was able to examine the British Library microfiches containing the preliminary part of the Eighteenth Century Short Title Catalogue (ESTC), although at this stage only the British Library's own holdings are listed.

Nearly all the 19th century German titles and editions are listed in Kayser's *Index Librorum*. There are some slight discrepancies between the actual title-page details and those given in Kayser. This is probably because the latter entries have been made mainly from publishers' catalogues and not from the actual volumes concerned. In some cases, there is a difference of a year either way in the date of the edition, no doubt for the same reason. In order therefore to suppress possible 'ghosts', only one date is given. The alternative date is given in the notes. When known, the title-page date is, of course, given preference.

All books in the author's collection are noted by 'CP'. The date in brackets after the reference name is that of the dates of the editions referred to. If no date is given every edition listed is given by the reference source. In the case of the bibliography given at the back of Brock's *A History of Fireworks*, there are a number of discrepancies between the date he gives and those shown in this bibliography. These are noted by a [?].

References and locations

Arber	ARBER, Prof. E. *The Term Catalogues 1668-1711.* London 1906
Backer	BACKER, Bibliothèque de la Compagnie de Jesus. Brussels 1890. (Ed. Sommervogel)
Barford	Lars Barfod collection (Denmark)
BC	Notified to the author by Bob Cardwell
Berlin	*Katalog der Architectur und Ornamentstichsammlung.* Berlin 1977
BL	*British Library Catalogue of Printed Books.* (Reading Room Edition)
BNC	*Bibliothèque Nationale. Catalogue des Livres imprimés.* Paris
Brinkman	BRINKMAN, Carel L. *Catalogus van Boeken 1933 onwards.* Amsterdam/Leiden
Brock	BROCK, Alan St.H. *A History of Fireworks.* London 1949
Brunet	BRUNET, J.C. *Manuel du Libraire et de l'Amateur de Livre.*
BU	*Bibliographie Universelle* (Michaud). Paris 1843-
Cockle	COCKLE, M.J.D. *A Bibliography of Military Books up to 1642.* London 1978
CP	Chris Philip Collection
CUBI	*Catalog Cumulativo. Biblioteca Nazionale Firenze (1886-1957).* Liechtenstein 1968
Du Pont	Du Pont collection. Eleutherian Mills Library, Delaware, USA
ECB	*English Catalogue of Books*
ESTC	*Eighteenth Century Short Title Catalogue.* (BL microfiche). 1983.

Fahler	FAHLER, E. *Feuerwerke des Barock.* J.B. Metzlersche, Stuttgart 1974
G&O	GEORGE, Carl & OST, Leopold. *Schlagvort Katalog.* Hannover 1889 etc
GV	*Gesamtverzeichnis des deutschsprachigen Schrifttums 1911-1965.* Munich 1977
Haan	HAAN, Bierens de. *Bibliographie Needlandesche Historique Scientifique.* Rome 1883
Hall	HALL, Trevor H. *A Bibliography of Books on Conjuring in English 1580-1850.* Palmyra Press, Lepton 1957
Hodgkin	HODGKIN, J.E. *Rariora.* London 1902
Jähns	JÄHNS, Max. *Geschichte der Kriegwissenschaften.* Leipzig 1889 (Reprint Olms 1966 3 vols.)
Kayser	*Kayser Index Librorum.* Leipzig 1834-1912
Lorenz	LORENZ, Otto. *Catalogue Generale Librairie Française depuis 1840.* Liechtenstein 1966-7
Lotz	LOTZ, A. *Das Feuerwerk.* Leipzig 1941
NBG	*Nouvelle Biographie Generale.* Dr. Hoefer, Paris 1858
NYPL	*Dictionary Catalogue of the Research Libraries* 1911-1971. New York Public Libraries–Research Libraries
NUC	*National Union Catalog.* Mansell, London
NUC SC	*National Union Catalog–Subject Index.* Various dates
Pagliani	PAGLIANI, *Catalogo Generale della Libreria Italiana.* (1847-1899)
Palau y Dulcet	PALAU y DULCET, *Manual del Librero Hispano-americano.* Barcelona 1948-73
Peddie	PEDDIE, R.A. *Subject Index of Books*
RAIL	*Royal Artillery Institution Library, Woolwich, London*
RDL	*Royal Danish Library, Copenhagen*

RDML	*Royal Danish Military Library, Copenhagen*
Rink	RINK, Evald. *Technical Americana. Checklist of Technical Publications Printed before 1831.* New York 1981
RL	ROWAN & LITTLEFIELD NUC 1956-1967
SOT	SOTHERAN, H. *Bibliotheca Chemica-Mathematica.* London 1921 etc
STC	POLLARD & REDGRAVE. *Short Title Catalogue (1475-1640).* Oxford University Press 1946
Stott	TOOLE-STOTT, R. *A Bibliography of English conjuring. 1581-1876* (2 vols.). Derby 1976/9
Vahl	VAHL, J. (and others). *Dansk Bogfortegnelse.* Copenhagen 1841 onwards
Wing	WING, D. *Short Title Catalogue (1641-1700).* Columbia University Press 1945-51 & 1972

Bibliography

A, J.L.y See CHERTIER, F. M

ACKALUSKY, James

Handbook for Making Rockets and Fireworks.

A010.1
Pageant Press: New York 1960
27p. illus. 21cms. RLA 011000 BL —

References: NUC Subject Catalogue. 1960

ADRIANUS, Romanus See ROMANUS, Adrianus

ALBERTI, Giuseppe Francesco Antonio (1712-1768)

La Pirotechnia o sia Trattato dei Fuochi d'Artificio. [*Pirotechnics: or a treatise on artificial fireworks.*]

A020.1
G. Battista Recurti: Venezia MDCCXLIX (1749)
TP+vi+128+21Pls.+ii(errori). 24cms. NA 0139502 BL 8715.ff.38

The first work in Italian to deal exclusively with the subject of fireworks for pleasure. The book claims 'to have modernised and improved the inventions of Frézier, Hanzelet, Henrion, Ozanam and Simienowicz'.

References: Du Pont. Berlin 3304. CP

ALEKSANDROV, N

Piroteckhnika. Feyerverochnoye delo. [*Pyrotechny and fireworks.*]

A030.1
Moscow 1930
NA — BL —

References: SHIDLOVSKY's bibliography.

AMATÖRFYRVERKAREN

Amatörfyrverkaren. Anvisningar till förfärdigande av moderna Fyrverkeripjäser. [*Amateur firework-maker. Instructions for the manufacture of modern firework pieces.*]

A040.1
Wahlström & Widstrand: Stockholm (1916)
PC+½T+TP+120(-6). 19cms. NA — BL —

References: Barfod

AMI(Y)OT, Joseph Marie (1718-1793)

Mémoires concernant l'histoire, des sciences ... des Chinois: par les Missionaires de Pekin. [*Vols. 7 & 8. Travels in China and Chinese Fireworks.*]

A050.1
Nyon: Paris 1782
xii+x+396+ Pls. NA 0283075 BL 146.b.5-19

Amiot was one of many 18th century proselytising Jesuit missionaries whose travels throughout Asia resulted in a great quantity of information about the Orient being published in the West. Much of the material in the above works comes from his translation of the *Wu Pei Chih* of 1628. A revised translation of some of the Firework material was made a few years earlier by another Jesuit missionary, Pierre d'INCARVILLE, part of which appeared in PERRINET d'Orval's *Manuel de l'Artificier*, published by Jombert in 1757. REINAUD and FAVE also reprinted INCARVILLE's text in their *Histoire de l'Artillerie*. Although Amiot's work is concerned primarily with the military aspects of Chinese pyrotechnics, it is an important work in helping to understand the development of fireworks in the Far East. A full description of Amiot's translation is given by PARTINGTON in his *History of Greek Fire and Gunpowder*(1960).

References: Backer

ANDERSON, N. D & BROWN, W

Fireworks! Pyrotechnics on Display.

A060.1
Dodd, Mead & Co.: New York 1983
79p. NA — BL —

An introduction to students in secondary schools. The treatment of fireworks is very superficial. Reviewed in Pyrotechnica Vol.IX.

ANDERSON, Robert (1688-1696)

The Making of Rockets ... Experimentally and Mathematically Demonstrated.

A070.1
Robert Morden: London 1696
TP+xiv+48. NA — BL 534.c.22

A070.2
Facsimile ed. (1961?)
48p. illus. 27cms. NA 0304458 BL —

The title page adds that the book is in two parts, 'The first containing the making of Rockets for the meanest capacity', the other,'to make Rockets by a Duplicate proposition, to 1000 pound Weight or higher'. The dedication is to 'the Rt. Hon. Henry, Earl of Romney ... Master General of his Majesty's Ordnance ... '. The book goes into great detail concerning the manufacture of rockets and is of a very practical nature. At the time of writing, the author had had 25 years experience in 'all sorts of great Artillery'

References: Brock (1696). Wing A3105

ANGELO, T

The Art of Making Fireworks by Plain and Easy Rules.

A080.1
J. Bailey: London c.1825
TP+FPl+28(-2). 17cms. NA 0324136 BL 1607/2116

A080.2
Second edition. J. Bailey: London c.1828
CoverTP+FPl+28(-2). 17cms. NA — BL

A080.3
(3rd ed.) J. Bailey: London c.1830
CoverTP+28(-2). 17cms. NSA 0062348 BL PO33015

The title page claims that Angelo was 'formerly of the Woolwich Laboratory', and the list of contents includes, 'directions for making Sky Rockets from 2oz. to 6lbs;

Tourbilions, Balloons & Mortars, Roman Candles, Stars etc.'; also 'Water Fireworks' and 'Waterloo Crackers' (Snaps). See also ANTONIO.

References: Du Pont (c.1820). Brock (c.1816). Barfod (2nd ed.)

ANON

The following list of *Anonymous* and *Pseudonymous* works can be found in alphabetical order under the name printed in capitals.

ANONYMOUS:

AMATÖRFYRVERKAREN anvisningar till förfädigande ..

ARMAMENTARIVM Principale

The **ART** of conjuring made easy ... With directions for making Fireworks

The **ART** of making Fireworks ...

A new and comprehensive edition of the **ART** of making and managing fireworks

The **BOOK** of Fireworks and wonders

BUCHSENMEISTEREI. Geschoss ...

Additamento ao VI. Volume da **COLLECÇÃO DE RECEITAS** e segredos particulares, ... de Fogo D'Artificio

The **COMPLETE** conjurer; with the whole art of making Fireworks

ENDLESS amusements ... in the art of making Fireworks

FIREWORKS and how to make them

The **GALLERY** or nature of Art. Pyrotechny or the Art of making Fireworks

Das **GANZE** der Feuerwerkerei ...

HANDBOK i Vatten-, Teater-, och Salongs-Fyrverkeri

HOCUS POCUS … to which is added … instructions for composing and displaying several most beautiful pieces of Fireworks

HOLIDAY frolics, containing … art of making Fireworks

THE JUVENILE BOOK FOR MAKING FIREWORKS — SEE, BAKER AND PARKS

KORT Afhandling om Lustfyrwerkeriet …

KORTFATTET anviisuing til at förfäerdige Lystfyrvaerkerier

Die **KUNST** der Feuerwerkerei

The **MAGIC** oracle … containing directions for making Fireworks

MANUALE del Pirotecnico

The **PERFECTION** of Military discipline

PIROTHECHNIE entretenida, curiosa y agredable de fuegos recreativos

PRACTICAL hints on Pyrotechny

PYROTECHNY, or the Art of making Fireworks …

SOULDIERS companion … together with … preparing Artificial Fireworks for war or recreation

The **TRUE** and easy guide to conjuring … to which is added the Art of making Fireworks

The **WHOLE** art of legerdermain … to which is added … the art of making Fireworks

The **WHOLE** secret laid open; or, the complete art of making the chemical fulminating objects …

The **YOUNG** man's book of amusements. Containing … the Art of making Fireworks

PSEUDONYMOUS:

L'Art de composer et faire les fusees volantes et non volantes
BAILLET DE SAINT-JULIEN, Baron Louis Guillaume

L'Art de faire a peu de frais les feux d'artifice
AUDOT, Louis Eustache

L'Arte di fare i fuochi d'artifizio ... di un Dilettante
DILETTANTE

The Art of making fireworks, improved to the modern practice ...
GROTZ, Christopher

The Compleat Gunner
VENN, Thomas

Fuegos Artificiales ... por J.L. y A
CHERTIER, F.M

Games for all seasons: consisting of ... Fireworks etc. ...
PARDON, George Frederick

Manuel de l'Artificier
PERRINET d'Orval, Jean Charles

Philosophical recreations ... a collection of experiments in ... Fireworks
BADCOCK, John

Practical hints on Pyrotechny
ELRINGTON, J.P

Pyrotechnia, of Konstige Vuurwerken ... door D.M
MANLYN, Daniel

La Pyrothecnie pratique, ou dialogues entre un amateur des feux d'artifice ...
GRIGNON & LALAING

TRACTATUS DE PYROTECHNIA -1766 MCZ

Traité des feux d'artifice ... par M. F * * * D.D.F.D.B
FREZIER, Amedée François

VOLLENKOMMENE vunterweisung ...
SYLVIUS, Nimrod

ANTONI, Alessandro Vittorio Papacino d'

Examen de la Poudre, traduit de l'Italien par M. le Vicomte de Flavigny.

A090.1
Amsterdam M.DCC.LXXIII (1773)
½T+TP+iv+240+9FPls. 20cms. NA — BL 1142.c.22

A Treatise on Gunpowder. A Treatise on Firearms and a Treatise on the Service of Artillery in time of War. Translated from the Italian of A.V.P. D'Antoni.

A090.2
Sold by T & J Egerton. London MDCCLXXXIX (1789)
½T+TP+xxxvi+374+ii+9FPls. 21cms. NA — BL —

Sotheran's note in his *Bibliotheca Chemico-Mathematica* says that, 'This remained for many years the standard work on its subject. It includes a large number of original experiments on Gunpowder, some of which anticipate the discoveries of Count Rutherford'. He wrote a number of other works which deal mainly with Artillery. I have not been able to trace any information concerning the original Italian work.

References: RAIL (1789)

ANTONI(J), Domenico

Trattato teorico, practico de Pirotecnia Civile. [*The theory and practice of recreative fireworks.*]

A100.1
Sambo: Trieste 1893
xv+283+12FPls. 23cms. NA — BL —

Provides one of the earliest detailed description of many types of 'Italian' (string-wound cylinder) style shells in use today including multi-break, parachute, comet shells, etc. It is very detailed and painstakingly illustrated. The Primo Catalogo Collettivo Delle Biblioteche Italiane (Rome 1967) give only one location, the Biblioteca Nazionale in Firenze.

References: Du Pont. Brock. Pagliaini. MS

ANTONIO, Carlos

The Art of Making Fireworks as Displayed at Vauxhall.

A110.1
J. Bushnell: London (c.1840)
Col.FPl+TP+24(-2). 19cms. NA — BL 8716.aaa.42

The title page informs us that ANTONIO was the 'Author of several Treatise on the Art of Fire-Works'. One has a feeling that he and ANGELO must either be the same person or he plagiarised. The coloured plate at the front has a strong family resemblance to that contained in ANGELO's work even to the boys shown dancing round a fire above which is suspended a hanging 'Guy'! Both books give details of making 'Waterloo Crackers' and 'Detonating Works' made with fulminate, a common pastime in the mid-19th century.

ANVIISNING

Anviisning til at forfaeridge Lystfyrvaekerier. [*Instructions for the production of recreative fireworks.*]

A120.1
H.P. Møller: Copenhagen 1852

NA — BL —

References: Vahl

ANWEISUNG

Anweisung kleine Lustfeuerwerke selbst v. mit wenigen Kosten verfertigen. [*Instruction for making fireworks cheaply for oneself.*]

A130.1
Stettin: Ulm 1811

NA — BL —

References: Kayser

APPIER, Jean (HANZELET LORRAIN) (1596-1647)

Recveil de Plusieurs Machines Militaires, et feux Artificials, pour la Guerre & Récréation. [*A description of many military machines, and artificial fireworks for war and recreation.*]

A140.1, A140.2
Charles Marchant: Pont-à-Mousson 1620
EngTP+xxii+88(-16)+TP+ii+40(-2)+TP+ii+24(-2)+TP+ii+112(-2)+ TP+ii+40(-2)+TP+ii+50(unum.)+TP+6. 20cms.

(a)	NA 0359607	BL 534.f.20.+
(b)	NA —	BL 1568/6428

Two versions of this exist. The first (a), entitled 'JEAN APPIER dit Hanzelet Calcographe et de FRANCOIS THYBOUREL Chyrurg.' and the second (b), 'FRANC.THYBOUREL. Maistre Chyrurgien et de JEAN APPIER dit Hanzelet et calcogr.' HODGKIN in his *Rariora* says that 'There was an evident jealousy between the joint authors. Thybourel the surgeon, employed by Appier, the engraver he tells us, and yet Appier managed as engraver to get his name first in the first issue, a proceeding evidently resented by his partner, and amended in the next. Ultimately, however, Appier dit Hanzelet got the upper hand, and his name alone appears in the much finer edition of 1630'. The book is divided into 7 'Livres', each with separate title-pages, except the first. They are subtitled as follows:- Book 2 - 'Deffencives', Book 3 - Untitled, Book 4 -'et de Fevx d'Artifice povr le guerre', Book 5 - 'et Fevx Artifices', Book 6 - 'Methode povr exrire occvltement a son amy par l'Alphabet De Trithemivs', Book 7 - 'Methode povr exrire de Nvict, a son Amy Absent, et Lvy Faire Concevoir Son Intention'. It is only on the first title page of version (a) that APPIER precedes THYBOUREL. On all other title pages THYBOUREL precedes APPIER.

A140.3
Charles Marchant: Pont-à-Mousson 1626
6 parts only. 20cms. NSA 0071529 BL —

La Pyrotechnie de Hanzelet Lorrain ou sont representez les plus rares & plus appreuuez secrets des machines & des feux artificials ... [*The pyrotechnics of Hanzelet Lorraine where are described the most rare and most learned secrets of machines and of fireworks.*]

A140.4
I.Gaspard Bernard: Pont-à-Mousson 1630
EngTP+vi+264illus. 24cms. NA 0359605 BL 534.i.4(3)

A140.5
Reprint. Editions des 4 Seigneurs: Grenoble 1971
½T+TP+viii+TP+vi+264illus. 21cms. NA — BL —

APPIER's influence on subsequent pyrotechnic works is immense. Trevor HALL's researches into the 17th century bibliography of *Mathematicall Recreations* describes the links between the *Recueil* of 1620, the first edition of ETTEN's *Recreations Mathematiques* to contain a section on Fireworks (Rouen 1628), APPIER's *La Pyrotechnie* of 1630, and that of Thomas MALTHUS and his French and English work of 1629. He further demonstrates that the Firework section of the English version of *Mathematicall Recreations,* normally ascribed to William OUGHTRED, was also probably made by MALTHUS. This book, in its turn, both in the French and English versions, fathered many similar works by authors such as OZANAM, HOOPER, GUYOT etc. The NUC Supplement lists a 1626 edition in which the title-page has had the date altered from 1620 with some misprints and errors corrected.

References: Du Pont (1620). Cockle 937, 938. Brock. Berlin 3295 (1629). CP (1630, 1971)

ARMAMENTARIVM

Armamentarivm Principale, oder Kriegsmunition vnd Artillerey-Buch. [*Principal armoury, or ammunition and artillery book.*]

A150.1
Johann Ammon: Franckfurt am Mayn 1625
EngTP+vi+113+v+illus+4Pls. 31cms. NA — BL —

Some three or four pages and illustrations are devoted to recreative fireworks, mainly wheels and mines. According to Max Jähns, Ammon edited and published what was, in effect, a shortened version of a manuscript by Helm which first appeared in 1536 under the title *Buch von den probirten Künsten.*

References: Cockle 690. RDML. Fahler. Jähns p995.

ART

The Art of Conjuring made easy; or, Instructions for Performing the Most Astonishing Sleight-of-Hand Feats. With directions for making Fireworks.

A160.1
T. Richardson: Derby c.1840
Col.Cover+12p. 14cms. NA — BL —

A160.2
Samuel and John Keys: Devonport c.1860
Col. Cover+12p. 17cms. NA 0437316 BL 7912.df.1/5

Apart from a small paragraph on 'Squibs and Serpents' and another on 'Detonating Balls & Waterloo Crackers', only the outside back-cover is devoted to 'Fancy Devices in Fireworks'. These consist of 'Golden Rain', 'Flower Pot', and a Dragon, made to move with a pulley, with serpents attached.

References: Stott 65/6. Hall 12

The Art of Making Fireworks and Fulminating Balls to which is added the History of Guy Fawkes and Gunpowder Plot...

A170.2
Second edition. W. Mason: London [1813]
PC+TP+28+1Pl. 19cms. NA — BL PO36045

A170.3
Third edition. W. Mason: London (1813)
1FPl+TP+26(-4). 18cms. NA — BL 8716.aa.6

A new and comprehensive edition of the art of making and managing fireworks with safety and ease, containing instructions for the following; artificial earthquakes, artificial thunder, blue flame ...

A170.4
Printed for T. Hughes: London c.1824
36p. 19cms. NA — BL —

References: NYPL

AUDOT, Louis Eustache (1783-1870)

L'Art de Faire a peu de Frais les Feux d'Artifice. [*The art of making firworks at little cost.*]

A180.1
Chez Audot: Paris 1818
TP+98+10Pls(Pl9F). 17cms. NA 0492270 BL PO46017

A180.2
2nd ed. Paris 1820
102+ Pls. NA 0492271 BL —

A180.3
3rd ed. Paris 1825
104p. 18cms. NA 0492272 BL —

L'Art de Faire a peu de Frais les Feux d'Artifice ... augmentée des nouveaux feux couleurs, des fusées a parachutes, et de notions sur la lumière électrique.

A180.4
Quatrième ed. L'Auteur: Paris 1853
½T+TP+104+10Pls.(1F)+xxivads. 20cms. NA 0492273 BL —

A180.5
Cinquième ed. Librairie Audot: Paris 1877
½T+TP+iv+164inc.86figs.+ivads. 20cms. NA — BL —

A180.6
Sixième ed. Librairie Audot: Paris 1891
PC+½T+TP+iv+196inc.128figs.+xiiads. 19cms.
NA — BL —

A popular and practical work that was much copied and translated. BUETTNER translated and published a German version in 1821 which went into several editions. CHERTIER's work on coloured fires is acknowledged in the 4th ed. and the use of the 'effets merveilleux de la lumière electrique' is commented on in the last two pages. The first edition is described as being 'Par un amateur', the second as 'Par M.L.E.A.'. It is possible that Audot was not the writer but only the editor, as he published a number of other technical 'how-to-do' works. Parts of this book were also included in other German works, together with a translation of some of CHERTIER's and RUGGIERI's work (q.v.). Another translation was made into Dutch by EISMA. In addition an Italian version was published by SONZOGNO, listed under DILETTANTE, which, although not a literal translation, nevertheless bears a strong family resemblance, both in the text and the plates.

References: Du Pont (1818, 1820). Brock (1818, 1823). BNC. CP (1853, 1877, 1891)

B, J See BADCOCK, J

B, T See TRATTATO TEORICO-PRACTICO

BABINGTON, John

Q922

Pyrotechnia, or a discovrse of Artificiall Fire-works: In which the true Grounds of that Art are plainly and perspiciously laid downe: Together with sundry such Motions, both Straight and Circular, performed by the help of Fire, as are not to be found in any other Discourse of this kind, Extant in any Language. Whereunto is annexed a short Treatise of Geometrie, ...

B010.1
Thomas Harper for Ralph Mab: London MDCXXXV (1635)
EngTP+TP+x+72+TP+ii+80+200inc.20Pls.(2F). 30cms.
NB 0006106 BL 1502/273

B010.2
De Capo Press, Theatrvm Orbis Terrarvm Ltd. Amsterdam 1971
½T+TP+EngTP+TP+x+72+20Pls.(2F). 31cms.
NB — BL 1653/295

Sotheran says: 'The first English work exclusively treating of Fireworks. It includes many elaborate conceits of the author'. HODGKIN in his *Rariora* considers it to be 'The most covetable, I think, of all English Books, at any rate on this subject ... The directions for making rockets, star-wheels and ground-wheels are more explicit than in earlier writers and evidence the experience of the master-gunner.' The 1971 reprint does not include the section on geometry. Allibone's entry in his *Critical Dictionary of English Literature* lists the edition as 1656 - presumably an error.

This book is without doubt the most important in Firework bibliography. It was the first work in English to deal exclusively with the subject of display fireworks, a fact often acknowledged in later works. All previous books which dealt with the subject concerned themselves primarily with the military uses of fireworks.

John Babington was an 'inferior Gunner' in the service of the Earl of Newport, Master of his Majesties Ordnance (Charles I). In his opening Epistle to his Master, he tells us that, 'I have bestowed some time and expence, as my occasions and ability would permit, in the practise of Artillery and Fire-workes, wherein though I have chiefly aymed at such conclusions as might be useful against an enemy in the field, yet these halcyon dayes of peace and tranquility ... have given mee occasion to invent and practise such also as these following; which howsoever they may seem to serve onely for delight and exercise, yet as by the handling of these there may bee gained knowledge in the natures and operations of the severall ingredients and their compositions; so the due consideration of the ordering of them may excite and stirre up in an ingenious minde, sundry inventions more serviceable in times of warre.' The emphasis has thus been reversed, and the peacetime uses of fireworks have now taken precedence over their wartime uses.

How much of the material contained in the book is truly original, and how much has been copied from previous publications? Babington's note 'To The Reader' starts as follows; 'Having for mine own private recreation, spent some vacant houres in the study and contemplation of those artificiall Fire-work which are discoursed in the ensuing Treatise, I held it requisite for my further and better satisfaction, as any opportunity did offer it selfe, to make experiment of some of the particularities there set downe; and I did accordingly now and then bring into Act that whereof before I had but a meere Notion, and the bare and naked Theorie: which because it could not be done privately, but that of necessity sometimes it would come to the view of others; by that means it has come to passe, that more notice than I was ever willing to, hath beene taken, that I have bestowed some thoughts upon that subject, and taken more than ordinary paines in it.'

It would therefore appear that Gunner Babington devoted some time to putting into practice what he writes about. Even so, one wonders where some of his more exotic ideas originated. A large part of the work is devoted to what might be termed 'pyrotechnic automata', that is to say, figures such as soldiers and dancers whose movement is derived from wheels driven by rocket motors. It can be no coincidence that one of the Eulogies printed in the introduction to Babington's work is from his, 'truly affectionate, and wel-wishing Friend, John Bate', whose *Mysteries of Nature and Art* had been published a year earlier 1634, by the same

Ralph Mab, the second edition of which, again published by Mab, appeared in the same year as Babington's work. It does not take much imagination to appreciate that Babington could easily have developed the use of rocket-motor power as a substitute for the water-power that Bate, in the first book of his work, uses to drive a number of automata - a singularly purposeless use of pyrotechnics which does not appear again in pyrotechnic literature, although animated lance-work set-pieces are still to be seen, especially in Malta, driven by gerbe propelled wheels. Nevertheless the scope and quality of plates in Babington far surpasses that of Bate.

Another source of inspiration to Babington would almost certainly have been the contemporary works on Artillery of which probably the greatest would have been those of FURTTENBACH whose *Halinitro-Pyrobolia* appeared in 1627. There are remarkable similarities between some of the plates and figures in the two books.

References: Du Pont. Brock. STC1099. CP (1635,1971)

[BADCOCK, John]

Philosophical Recreations, or, Winter Amusements: a Collection of Entertaining and Surprising Experiments in mechanics, arithmetic, optics, hydrostatics, hydraulics, pneumatics, electricity, chemistry, magnetism, & pyrotechny, or, Art of Making Fire-Works ... The whole simplified and clearly elucidated so as to suit every capacity.

B020.1
For Thomas Hughes: London [1820]
ColFP+TP+xiv+200. 15cms. NB 0024305 BL 8705.aa.31

There is no author's name on the title-page, but the preface is signed 'J.B.'. The date of publication is given in his *Domestic Amusements ... a Sequel Volume to Philosophical Recreations* which does not contain any material on fireworks. Pages 194-200 are devoted to 'The Art of Making Fireworks', including gunpowder, touch-paper, scrole wheels and suitable fire for them, portfires, sparkes, crackers, squibs and serpents. Stott gives two other variant imprints.

References: Hall 18-19. Stott 75-77

[BAILLET DE SAINT-JULIEN, Baron Louis Guillaume] (c.1715-c.1775)

L'Art de composer et faire les fusées volantes et non volantes. [*The art of making flying rockets and line rockets.*]

B030.1
(1st ed.) Paris 1775
NB — BL —

B030.2
Seconde Edition. Chez d'Houry: A la Haye M.DCC.LXXVI (1776)
TP+44(-2)+1FPl. 20cms. NB — BL —

B030.3
(3rd ed.) Chez Nyon: Paris M.DCC.LXXX (1780)
TP+42+1FPl. 20cms. NB — BL 1236/1560

NBG. 'N.T. Le Moyne des Essarts - *Les siecles litteraires de la France*. Paris 1800'. The title-page of the 3rd edition adds 'Par l'auteur de la Maniere d'enluminer l'estampe posee sur toile'.

References: Barbier. CP (1776)

BARRY, Gerat

By Sea and Land. A discourse of Military Discipline devided into three boockes ... The laste booke treatinge of Fire-wourckes of rare executiones by Sea and Land.

B040.1
Widowe of John Mommart: Bruxelles M.DC.XXXIV (1634)
EngTP+TP+vi+211+xvii+10Pls.+FTs. 28cms.
NB 0151650 BL 719.L.17

The third book deals only with Fire-tubes, Fire-balls, Arrows and Pots. There is no mention of any recreative use

References: STC1528. Cockle 130

BATE, John

The Mysteryes of Natvre and Art. Conteined in foure severall Tretises, The first of Water-Workes, The Second of FYER Workes, the Third of Drawing, Colouring, Painting and Engraving. The fourth of divers Experiments, as well serviceable as delightful: partly Collected, and partly of the Authors Peculiar Practice and Invention by I.B.

B050.1
Imprinted for Ralph Mab: (London) 1634
EngTP+viii+192(incl.7Pls). 19cms. NB 0180098 BL C.27.h.5

The Mysteries of Nature and Art. In foure severall parts. The first of Water Works. The second of Fire Works. The third of Drawing, Washing, Limning, Painting and Engraving. The fourth of Sundry Experiments. The second edition with many additions unto every part. By Iohn Bate.

B050.2
Printed for Ralph Mab: (London) 1635
FP(Bate)+EngTP+x+288+xvi+Pls. 19cms. NB 0180102 BL C.122.e.18+

The Mysteries of Natvre and Art. In four severall Parts. The First of VVater-VVorks. The Second of Fire-VVorks, The Third of Drawing, Colouring, Limming, Paynting, Graving, and Etching. The Fourth of Experiments. By John Bate.

B050.3
(3rd ed). R. Bishop for Andrew Crook: London 1654
EngTP+TP+iv+230(-4)+viads. 19cms. NB 0180105 BL 1044.i.27

The second Booke. Teaching most plainly, and withall most exactly, the composing of all manner of Fire-works for Tryumph and Recreation.

B050.4
(Reprint) Shorey Book Store: Seattle 1972
NB — BL —

Although each edition consists of four parts and the paging is continuous, each part has its own title-page. The title-page of the Second Book of the first edition is, 'Teaching most plainly, and withall most exactly, the composing of all manner of Fire-works for Triumph and Recreation'. The second edition has 'Tryumph', whilst the third has 'The composing of all manner of Fier-Works for Triumph & Recreacion, plainly and easily taught by JOHN BATE'. The reprint of the 'Second Booke' of 1635 was reviewed in the January 1972 issue of *American Pyrotechnists Fireworks News*.

In his note 'To the Reader' at the beginning of the Second Book (1634 edition; slightly altered in the 1635 edition) Bate mentions that, 'there hath a desistance been occasioned since the inception of this worke, by reason of the occurents of certain Authors, that contrary unto my knowledge had laboured so fully herein; but after consideration had (that for the most part they were but translations) I thought it might be no lesse lawfull and commendable for mee than for others, to communicate unto such as are yet desirous of further information, that wherein I have bestowed such cost and paines.'

It is possible that he is refering to his friend BABINGTON, but it is more likely that he refers to the publication in 1633 of the English translation (sometimes attributed to Oughtred) of Van Etten's *Recreations Mathematiques*, which contains a substantial section on Fireworks, similar in many respects to his own. Indeed one of Bate's paragraphs is headed 'Composition of Rockets of all sizes, according unto the prescription of the noted Professors, as Mr. Malthus, Mr. Norton, and the French Author, Des Recreationes Mathematiques.'

A detailed collation of all three editions is described by Prof. John Ferguson in his *Bibliographic Notes on the Histories of Inventions and Book of Secrets*: Glasgow 1895.

References: STC1577-8. Wing B1092. Du Pont (1654). Brock (1635). Stott 81-83. Hall 20-22. CP (1635)

BEBIE, Jules

Manual of Explosives, Military Pyrotechnics and Chemical Warfare Agents.

B060.1
Macmillan & Co.: New York 1943
xiii+171p. NB 0231501 BL 8899.f.24

B060.2
(Reprint) Paladin Press: Colorado 1977
xiii+171p. NB — BL —

The introduction states the aim of the book is, 'to be of service to scientific and technical workers in the field of explosives and war chemicals.' It is essentially a Dictionary of Chemicals and Explosives for military use and the bibliography only relates to the chemistry of military explosives, although of course some materials are common to fireworks as well. The definitions are brief and simple.

BEGREB

Begreb, kort, or Lystfyrvaerkerier, for Lystfyvaekere. [*Brief treatise on fireworks for celebrations.*]

B070.1
H.P. Møller: Copenhagen 1861
NB — BL —

Probably by NÆSER

References: Vahl

BELIDOR, Bernard Forest de (1697-1761)

Le Bombardier François ou Nouvelle Methode de Jetter les Bombes avec Précision. [*The French bombardier, or new methods of firing bombs with precision.*]

B080.1
De l'Imprimerie Royale: Paris M.DCCXXXI. (1731)
Eng+TP+iv+TP+xlii+TP+272+TP+(275-366)+8FPls. 25cms.
NB 0275605 BL 62.e.9

B080.2
Aux Depens de la Compagnie: Amsterdam M.DCC.XXXIV (1734)
Eng+TP(BR)+iv+TP+xlii+TP+272+TP+(275-366)+8FPls. 27cms.
NB 0275606 BL 1572/348

Französcher Bombardier oder neue Art die Bomben gewiss zu werfen nebst einer Abhandlung von Ernst- und Lust- Feuer-Werken. Aus dem Französischen in Teutsche gebracht.

B080.3
C. Weigels: Nürnberg 1756
EngTP+TP+ NB — BL —

Belidor was Professor of the Artillery school at La Ferté in 1758 and then became Director of the Arsenal and Inspector of the Corps of Miners. He was essentially a very practical gunner as his detailed firing tables show. The first part deals with the firing of bombs followed by 272 pages of tables of elevation. The second part of the book, from page 273 onwards, deals with Gunpowder, 'Feux d'artifice pour la Guerre' and the last 54 pages are devoted to the 'Traité des feux d'artifice les plus usitez dans les rejouissances.' Six of the plates relate to Fireworks.

References: Jähns. CP (1731)

BERENDSOHN, B.S See LEBLANC, Henri

BIBILE BANDARA

Practical Firework Making. (Translation of Sinhalese pamphlet.)

B090.1
Columbian Press: (Colombo) 1897
6p. NB — BL 14165.d.19/2

BINNING, Capt. Thomas (1620- ?)

A light to the Art of Gunnery ... likewise the ingredients and making of the most necessary Fireworks.

B100.1
J. Darby for the Author: London 1676
(Eds. 1677, 1689, 1703)

On page 154 of the first edition the author has this to say on the subject of recreative fireworks, 'There are several sorts of Fireworks, some for offensive service, some for defensive service, and some for recreation and sport: I intend only to speak of those which are to be used in earnest, not minding to meddle with those for Recreation at this time, in regard they are so learnedly treated by divers authors.'

References: Wing B2934B-E

BIRINGUCCIO, Vannocio (1480-1539)

De la Pirotechnia. Libri X. ...

B110.1
Venturino Roffinello: Venetia M.CCCCC.XL.(1540)
viii+168L. 22cms. NB 0501065 BL 717.g.36

HM 335/£2000 D267/£3200 D280 C/1/4/81/£3000 JN5

Pirotechnia. Li diece libri della Pirotechnia, ...

B110.2
Giouan Padoano: Venetia M D L.(1550)
viii+168L. 22cms. NB 0501071 BL 8715.dd.15

MS20/£1100 D262

B110.3
C.da Trino di Monteferrato: Vinegia M D L VIII(1558)
viii+168L. 22cms. NB 0501072 BL 990.d.2

HM 336 £550

Pirotechnia del S. Vannvccio Biringvccio Senese; ...

B110.4
P. Gironimo Giglio: Venetia M.D.LIX.(1559)
TP+345(-1)L+xiv+WC. 16cms. NB 0501074 BL 1033.3.20

Pirotechnia del Signor Vannvccio Biringvccio Senese, ...

B110.5
G. Longhi: Bologna 1678
630p. 17cms. NB 0501075 BL 957.b.37

D267/£220

La Pyrotechnie, ov art dv fev, contenant dix livres ... Composee par le seigneur Vanoccio Biringuccio Siennois. Et traduite d'Italien en Francois, par feu maistre Iaques Vincent ...

B110.6
Claude Fremy: Paris 1556
230+4Pls. 23cms. NB 0501080 BL 1033.k.1

B110.7
Claude Fremy: Paris 1572
168+4Pls. 22cms. NB 0501081 BL 953.f.1

B110.8
I. Cailloüé: Rouen M.DC.XXVII.(1627)
228+2Pls. 22cms. NB 0501084 BL 457.c.12

B110.9
J. Correard: Paris 1856
124+2Pls. 24cms. NB 0501085 BL —

Biringuccios Pirotechnia.

B110.10
Friedr. Vieweg & Sohn: Braunschweig. (1925)
544+ 22cms. NB 0501076 BL —

The Pirotechnia of Vannoccio Biringuccio. Translated from the Italian with an introduction and notes by Cyril Stanley Smith & Martha Teach Gnudi.

B110.11
Basic Books Inc.: New York 1942
NB 0501078 BL —

B110.12
Basic Books Inc.: New York 1943.
NB 0501079 BL —

B110.13
Basic Books Inc.: New York 1959
½T+TP+xxiii+TP+TP+478. 28cms. NB — BL —

B110.14
(Reprint) Massachusetts Institute of Technology: 1960
NB — BL —

This is a much-quoted work although the title is somewhat misleading. In 16th-century Italian it meant metallurgy. Only Chapter X deals with 'Artificial Combustible materials ... commonly called Fireworks to be used in offensive and defensive warfare and for Festivities on Holidays'. The eleven chapters are titled as follows:- (I) 'Concerning the nature of Saltpeter ...' (II) 'Concerning the Powder used for Guns ...' (III) 'Concerning the methods that are used for loading Guns ...' (IV) 'Concerning Mines ...' (V) 'How Fire-tubes should be made for defending or assaulting Batteries or Gates, for burning Supplies, or for Festivals.' (VI) 'The Method of Making Metal Balls ...' (VII) 'The Method of making tongues of Fire ...' (VIII) 'The method of preparing Fire Pots ...' (IX) 'The manner of compounding various Incendiary compositions which are commonly called Fireworks'. (X) 'Method of preparing Fireworks called Girandoles, which were once customarily used in some Tuscan cities for magnificent Displays for Public Festivities on Solomn Feast Days.' (XI) 'Concerning the Fire that consumes without leaving Ashes ...'. The solution to this riddle is 'LOVE' - shades of Turandot!

References: Du Pont. Cockle 931. Brock (1540). Berlin 3293 (1540). CP (1559, 1959)

BLONDEL, C

Der Selbstlehrende Feuerwerker oder gründliche Anweisung zur Lustfeuerwerkskunst für Liebhaber. [*Teach yourself fireworks or thorough instructions for lovers of the art.*]

B120.1
Joachimsch: Leipzig [1802]
TP+xiv+175+5FPls. 20cms. NB — BL PW45/44946

B120.2
Neueste Auf. np.: Leipzig 1803
TP+iv+190+3Pls. 18cms. NB — BL

This contains an interesting bibliography of some 35 16th- and 17th-century artillery and firework books.

References: Kayser. Barfod (1803)

BLUEMEL, Johann Daniel

Deutliche und Gründliche Anweisung zur Lust= Feuerwerkery ... [*Clear and basic instructions for recreative fireworks.*]

B130.1
A. König: Strasburg 1765
TP+ii+140+iv+9FPls. 21cms. NB — BL 1142.e.29

Grundliche Anweisung zur Lust Feuerwerkery . . . .

B130.2
A. König: Strasburg 1771
TP+180+10FPls. 21cms. NB 0562179 BL M.L.ccc.22

The first 9 plates are common to both editions. The 1771 edition contains a supplement by Michael Höckely which gives the first printed description of the pinwheel, that is a spirally wound tube filled with composition that drives it round.

References: Du Pont. Berlin 3306. Kayser (1771). CP (1765, 1771)

BOOK

The Book of Fireworks and Wonders.

B140.1
Turner and Fisher: New York c.1965
36p. 16cms. NB — BL —

References: NUC SC 1965

BOTTEE de TOULMONT, Jean Joseph Auguste (1764-1816) and RIFFAULT

Traité de l'Art de Fabriquer la poudre à canon ... Accompagné d'un Recueil de quarante plaches au traite. [*A treatise on the art of making gunpowder.*]

B150.1
Leblanc: Paris 1811
TP+viii+clxiii+TP+9Tbls.+537. 27cms. NB 0681905 BL 59.b.16

B150.2
(Plates) Leblanc: Paris 1811
TP+39Pls. 27x52cms. NB — BL 3.TAB.44

This large and sumptuous work deals very fully with the history and manufacture of gunpowder in France.

BOURNE, William (? -1583)

Inuentions and Deuices ...

B160.1
Thomas Woodcock: London 1578
EngTP+xxii+100(inc.4figs.+2Tab.). 17cms. NB 0700062 BL C.71.cc.16+

Contains 113 'Devices, very necessary for Generalles and Captains, or Leaders of men, as well as by Sea as by Land'. None of them appear to have any bearing on fireworks although BROCK in his *History of Fireworks* quotes the following passage, 'Divers gunners and other men have devised sundry sorts of Fireworks for the annoyance of their enemies, yet as far as I have ever seen or heard, I never knew any good service done by it, either by sea or land, but only by powder ... But for their other fireworks it is rather meet to be used in the time of pleasure in the night rather than for any service.' In 1587 Bourne first published a work entitled *The Arte of Shooting in great Ordnaunce ...*; another edition of this, published in 1643, contained, in addition, Robert NORTON's *The Gunner's Dialogue*. They were clearly expected to be read together since they are paged continuously.

References: Cockle 24 & 35. Brock. STC3421

BOVY, Johan (de Liege)

La Pyrotechnie Militaire 1591. Extrait du *Memorial de l'Artillerie de la Marine.*

B170.1
Berger-Levrault: Paris 1892
PC+½T+TP+iv+68(-4)inc.39fig. 22cms. NB — BL 8830.g.33/6

The title-page adds 'La Pyrotechnie Militaire, redigee en 1591 par M. Johan Bovy de Liege, provient d'un manuscrit inedit de la B.N.'. The figures that accompany the text, 'reproduites par la photographie afin de leur conserver toute leur originalite' include some remarkably life-like cross-sections of rockets and 'trois roues pour feux de joye'. They seem more suited to peace than war!

References: BNC

BRANTZIUS, Ioannes

Artifice de Feu, & Diuers Instuments de Guerre. Das ist, Kunstlich Feurwerck und Kriegs Instrumenta ... [*Fireworks and other instruments of war*]

B180.1
A.Bertram: Strasburg 1603
EngTP+vi+184inc.91Pls. 31cms. NB — BL 1505/180+

This is a German translation of BOILLOT's *D'Artifices de Feu*. The 1603 edition contains an extra page of text and plate not in the 1598 original. Hodgkin has this to say; 'A treatise printed in German and French in parallel columns of text, with 91 etchings by J. Boillet, a moderate draughtsman, and evidently a mere tyro in the management of his acid-bath and the other details of the etching process. The resulting plates are however not devoid of a certain quaintness. The engines of various constructions, which were to do rude damage to the gates of the enemy, might possibly have indented or even shattered a feeble portal, if only the besieged were good-natured enough to allow an interval of twelve hours for their precise attachment, but to us they can only appear frivolous in the extreme ... The invention of gun-powder obtains a passing notice, and Schwartz comes in for recognition as a claimant of the honour of the discovery ... The recreative portion contains illustrations of a considerable variety of balls for use on sea and land and anticipations of several of the schemes claimed as new by Hanzelet (APPIER) and Thybourel sixteen years later'.

References: Cockle 933

BRAUER, Karl O

Handbook of Pyrotechnics.

B190.1
Chemical Publishing Co.: New York 1974
ix+402 22cms. NUC 75-33937 BL —

Despite its title, it deals in the main with explosive-actuated devices such as bolts, line-cutters , propellants, time-delay compositions and the like. Only of interest to recreative pyrotechnists, are the chapters on electric igniters, flares and smoke generators. It was reviewed in the June 1974 edition of *American Pyrotechnist Firework News*.

BRAUN, Ernst

Novissimum Fundamentum & Praxis Artilleriae ... Ferner eine vollkommene Beschreibung der Lust-Sachen und Haupte-Feuerwercke auf dem Lande und Wasser wie solche zu arbeiten und vor Könige und Fürsten darzustellen sind ... [*The newest theory and practice of artillery ... furthermore a complete description of the entertainments and principle fireworks on land and water, how these are to be made and shown before Kings and Princes.*]

B200.1
J.F. Gräfen: Danzig 1682
EngTP+TP+ii+197+viii+Pls. 32cms. NB 0754737 BL —

B200.2
J.Z. Stollen: Danzig 1687
EngTP+TP(BR)+vi+198+ii+26Pls. 33cms. NB — BL M.L.p.13.+

References: RDML (1687). Fahler (1682). Lotz (1682). RAI (1682)

BRAUN, Wernher von. & ORDWAY III, F. I

The Rockets' Red Glare.

B210.1
Anchor Press: New York 1976
½T+TP+vi+212illus. 21cms. NB — BL —

References: CP (1976)

BRECHTEL, Frantz Joachim

Büchsenmeisterey, Das ist; Kurtze doch eigentliche erklerung deren ding so einem Büchsenmeister fürnemlich zu wissen von nöten ... Sampt getreulicher vnterweisung mancherley Feuerwerck wie derselben eins theils zum ernst anderstheils aber zum schimff vnnd lust bereittet zugericht vnnd gebraucht werden sollen. [*Artillery, a short but essential explanation which a gunner should know ... with reliable instructions for many kinds of fireworks, one part for serious use and another for gay and joyous use.*]

B220.1
K. Gerlachin: Nürnberg M.D.XCI. (1591)
TP(BR)+336inc.WC 16cms NB — BL 1438.b.13/1

B220.2
Paul Kauffmann: Nürnberg 1599
TP(BR)+336inc.WC 16cms NB — BL —

De Bussen Meesterye ... an allerhande geschut te schieten en de van salpeter poeyer ende alderhande Vyerwercke sonderlich te bereyden. [*(Dutch edition) Gunnery ... to shoot all kinds of ordnance and to make saltpeter powder and various fireworks.*]

B220.3-7
(Amsterdam 1591, 1594, 1605, 1625 & 1630)

Contains a chapter describing how to construct a castle firework machine. According to Jähns SCHMIDLAP's work, *Künstliche und rechtschaffene Fewrwerck*, was added to the 1625 Amsterdam edition.

References: Du Pont (1599). Cockle 667 (1591, 1599). Fahler (1591). Lotz (1591, 1625, 1630). Jähns p652

BROCK, Alan St.Hill (1886-1956)

Pyrotechnics: The History and Art of Firework Making.

B230.1
Daniel O'Connor: London MCMXXII (1922)
xvi+198+36Pls. 25cms. NB 0826430 BL 8716.ccc.4

A History of Fireworks.

B230.2
George Harrap: London 1949
ii+Col.Pl+TP+ii+280(-6). 25cms. NB 0826428 BL 8715.i.47

BROCK's works are unique in the literature of fireworks since they are the only ones to provide an overall worldwide history of the subject. Starting with the discovery of gunpowder and the uses of fireworks for celebrations and ceremonies down the centuries, they also cover the manufacture of the various items and the history of their introduction into the art. Although he naturally emphasises the family's own considerable influence on the industry throughout the last two or three hundred years, nevertheless he provides indispensable reading for layman and professional alike. His far from complete bibliogaphy has provided virtually the only source of information on the subject until now. The NUC under NB 0826431 gives a 1927 edition. I suspect this may be a ghost.

BROCK, Alan St. Hill & MURRAY, Webster

Fireworks & Fetes.

B240.1
Penguin Books: Harmondsworth & New York (1946)
31p inc.col illus. 18x22cms. NB — BL WP6717/62

This is a Puffin Picture Book, a series that was angled towards children. Full of facts about fireworks and displays with many coloured drawings.

References: CP (1922, 1949, 1946).

BROWNE, Dr. William Henry

The Art of Pyrotechny: being comprehensive and practical instructions for the Manufacture of Fireworks, specially designed for the use of Amateurs.

B250.1
'The Bazaar', (By Frederick Phillips) London c.1873
Pl+TP+ii+148(-6)170figs.+xviads. 21cms. NB — BL

B250.2
'The Bazaar', (By Alfred Bradley) London [1880]
Pl+TP+ii+148(-6)170figs.+xviads. 21cms. NB 0873823 BL 8908.aa.25

Minor Fireworks: Containing Contributions for the manufacture of the common and simple varieties of Firework for the use of Amateurs.

B250.3
'The Bazaar' :London [1878]
PC+46+ii+2Pls(illus.). 19cms. NB 0873825 BL 1147.e.16/5

Firework Accidents.

B250.4
W. Adams: Hull 1884
NB — BL —

Firework Making for Amateurs: being Complete and Explicit Instructions in the Art of Pyrotechny. Especially prepared for the use of Amateurs and Beginners.

B250.5
L. Upcott Hill: London c.1888
½T+TP+iv+200illus.+ads. 19cms. NB 0873824 BL —

Dr.Browne was probably the best of the English pyrotechnists writing during the 1870-1890 period. The inside back cover of *Minor Fireworks* tells us that Browne was an Analytical, Consulting & Pyrotechnic Chemist from Aldborough, Hull, and that he had well equipped laboratories to carry out 'Investigation of Explosive Compounds' as well as the preparation of chemicals. He also gave 'Private lessons at pupils own residences when required, with personal superintendence of their Displays'. There are two versions of *The Art of Pyrotechny*, the first was published from 32 Wellington Street, and the second from 170 Strand, London. The two issues appear identical, including misprints, except that the type size is larger on the Plates of the second issue. *Minor Fireworks* was issued as a sixpenny pamphlet in a paper cover and contains details of how to make squibs, blue candles, star lights, pin wheels, crackers, maroons, saucissons, parlour fireworks and the like. In the introduction to *The Art of Pyrotechny* he mentions that, 'the simplest and best (work on pyrotechny) that has been published of late years is one printed by Wm. Underwood, Eden-quay, Dublin, in 1850, now out of print.' For a long time I was unable to establish what work this might be; however when examining the firework books held by the Patent Office Library (now part of the Science Reference Library of the British Library), I found a book called *Practical Hints on Pyrotechny* printed anonymously by Underwood's of Dublin in 1850. On the title-page was a note to the effect that it was given to the Library by 'the author, J.F. Elrington'.

References: CP (1880, 1888). Brock (1880, 1884)

BRUCE, S

A New Guide to Pyrotechny.

B260.1

1840

NB — BL —

There was a copy in the Patent Office Library, London (now part of the British Library), but unfortunately it is missing.

References: Peddie

BRUHL, Frederick

The Art of Making Fireworks.

B270.1
W. Brittain: London 1844
TP+iv+66(60eng.) 19cms. NB — BL 8715.aa.47/3

Similar to many other firework 'How-to-do' books of the period, although it does describe some rather more complex set-pieces than most. Potassium Chlorate and metal salts are used for the colour compositions. Contains chapters on fire balloons and fulminate experiments. Also 'Artificial Fireworks', i.e. Firework effects obtained by using transparent paper scrolls.

References: `Brock

BRY, Johann Theodor de (1561-1623?). See also **UFANO, Diego**

Kunstbuchlein von Geschutz und Feuerwerk .... [*Booklet of the art of guns and fireworks.*]

B280.1
P. Jacobi: Frankfurt am Mayn 1619
TP+viii+36+2Pls. 29cms. NB 0901168 BL 8715.k.4

References: Cockle 686 et.al.

BUCHNER(N), Johann Siegmund

Theoria et Praxis Artilleriae. [*The theory and practice of Artillery. (In three parts with separate titlepages.)*]

B290.1
Christian Sigmund Froberg: Nürnberg 1682-1695
(I)EngTP+TP+x+88+iv+23Pls+(II)EngTP+TP+iv+94+vi+16Pls+ (III)EngTP+TP+56+9Pls. 32cms.
NB 0913713 BL 8825.f.34+

Part One deals with Artillery, Part Two with Fireworks and Part Three with Gunpowder.

References: Fahler. RDML (1682/83/85). Barfod (1682/90/95). RAI (1685/83/83). Jähns.

BUCHSENMEISTEREI

Büchsenmeisterei. Geschoss, Büchsen, Pulver, Kugeln, Salpeter, Feurwerck vnnd Pfeilöc. Zum Schimff und ernst zumachen züzurichten vnnd nach jedes Gewicht Stein und Lot zugebrauchen. Dabei Gemeine Kreigsrecht Räth Regiment vnd Ordnung. [*Gunnery. Cannon, guns, powder, balls, fireworks and saltpeter. How to make and use for recreational and serious purposes ...*]

B300.1
Christian Egenolffs Erben: Franckfurt am Mayn 1559.
TP(BR)+95L 15cms. NB — BL —

References: RDML

BUETTNER, F. C. A

Die Feuerwerkerkunst für Dilettanten, oder Anleitung, wie man mit wenig Kosten alle Arten von Kunst= und Lustfeuern, zur Berschönerung häuslicher Feste, zubereiten kann. Aus dem Französischen von F.C.A Büttner [*The art of fireworks for amateurs, or instructions on how to make at little cost all kinds of artistic and recreative fireworks.*]

B310.1
B.F. Voigt: Sondersh 1821
NB — BL —

B310.2
2e. verb. u verm. aufl. B.F. Voigt: Ilmenau 1823
TP+56+10Pls. 18cms. NB 0934867 BL —

B310.3
3e. aufl. B.F. Voigt: Ilmenau 1829
56+10Pls. 18cms. NB — BL —

B310.4
4e. verb. u. verm. aufl. B.F. Voigt: Ilmenau 1834
56+10Pls. 18cms. NB — BL —

Die Lustfeuerwerkerei für Dilettanten ....

B310.5
5e. sehr verb. aufl. B.F. Voigt: Weimar 1837
NB — BL —

Der Excellirende Lustfeuerwerker ... aus dem Franzosischen des F.M. Chertier. [*The skilled fireworker ... from the French of F.M. Chertier.*]

B310.6
6e. aufl. B.F. Voigt: Weimar 1837
PC+vi+70+11Pls(1FPl). 19cms. NB — BL 896.dd.37

Die Lustfeuerwerkerei für Dilettanten. (1st Pt.) Aus dem Franzoisischen von F.C.A. Buettner. (6e sehr verb. aufl.) (2nd Pt.) Enthaltend den excellirenden Lustfeuerwerker. Aus der Franzoisischen des F.M.Chertier von H.Gauss.

B310.7
B.F. Voigt: Weimar 1841
NB — BL —

Die Lustfeuerwerkerei für Feuerwerker und Dilettanten ... von F.C.A. Büttner und F.M. Chertier ... Neu bearbeitet von Heinr. Hamberger.

B310.8
7e verm. u. verb. aufl. B.F. Voigt: 1846
xiv+144+12Pls. 18cms. NB 0934868 BL —

Die Lustfeuerwerkerei für Feuerwerker und Dilettanten ... von F.C.A. Büttner, F.M. Chertier, H. Hamberger ... von Fr. Harzer.

B310.9
8e sehr verm. u. verb. aufl. B.F. Voigt: Weimar 1857
½T+TP+viii+173+viads.+12Pls(inc.2FPls). 19cms.
NB — BL PO6643

Die Lustfeuerwerkerei zur Verschönerung öffentlicher und hauslicher Feste ... von F.C.A. Büttner, F.M. Chertier, H. Hamberger, und Fr. Harzer ... herausgegeben von Ferdinand Hottinger. [*Making fireworks for embellishing public celebrations.*]

B310.10
9e verm. u. verb. aufl. B.F. Voigt: Weimar 1864
PC+½T+TP+viii+245(55illus.)+3Pls. 19cms.
NB 0934869 BL —

DUTCH EDITIONS

De vervaardiging van kunstvuurwerken, ter opluistering van openbare en bijzondere feesten ... In de vroegere uitgaven bewerkt door F.C.A. Büttner, F.M. Chertier, H. Hamberger en Fr. Hartzer. 9e. verm. en verb. uitgave. [*The manufacture of artificial fireworks for enhancing public and special celebrations.*

B310.11
J.B. Peeters: Gouda 1867
NB — BL —

B310.12
P. Kluitman: Alkmaar 1886
NB — BL —

Buettner as editor and translator, and Voigt as publisher issued a large number of magazines, leaflets, and technical pamphlets of all kinds. The 1821 edition of *Feuerwerkerkunst* was a translation of AUDOT's *L'Art de faire a peu de frais les feux d'artifice*. The '6th edition' of 1837 includes a German translation of CHERTIER's *Essai* originally published in 1836. From 1841 these editions were published under the general title of *Neuer Schauplatz der Kunst und Handwerken* (New Presentation of Arts and Crafts) No.92. Details of all nine editions are to be found in Kayser. The 10th and 11th editions were edited by SANDMANN and are listed under this name by Kayser and other Bibliographies.

References: Du Pont (1823). Kayser (German editions). Brinkman (1867, 1886). CP (1823, 1864)

BUJARD, Alfons (1861-1917)

Leitfaden der Pyrotechnik. Einfuhrung in die Chemie der wichtigsten Rohmaterialien und Sprengstoffe der Kunstfeuerwerkerei, nebst Besprechung der einschlagigen Polizei- und Eisenbahnverkehrsverordnungen ... [*Guide to pyrotechnics. Introduction to the chemistry of the most important raw materials and explosives for making fireworks, together with a discussion on the relevant police and railway traffic regulations ...*]

B320.1
A. Bergstrasser: Stuttgart 1899
x+260+illus. 24cms. NB 0941288 BL —

Die Feuerwerkerei. [*The making of fireworks.*]

B320.2
G.J. Göschen'sche: Berlin und Leipzig 1912
iiiads.+TP+88inc.6figs.+32p.cat. 15cms. NB — BL —

References: Du Pont. G&O (1912). CP (1912). Brock (1912)

CAPOBIANCO, Alessandro

Corona e palma Militare di Artegliera ... fuochi artificiati da Giuoco, e Guerra ...

C010.1
Gio. Antonio Rampazetto: Venetia 1598
EngTP+vi+58Linc.WC. 33cms. NC 0120465 BL —

C010.2
Francesco Bariletti: Venetia M DCII(1602)
EngTP+vi+58Linc.WC. 33cms. NC 0120467 BL 534.m.1

C010.3
1618
NC — BL —

C010.4
Antonio Bariletti: Venetia M DCXXXXVII (1647)
EngTP+iv+60Linc.WC. 33cms. NC — BL —

Some of the leaves are misnumbered (26 numbered 29, 52 numbered 48 and 58 numbered 85). In the 1647 edition these have been corrected and two additional leaves have been included, entitled 'Essercitio Militare'. The section entitled 'Dell'Inuentioni de fuochi artificiati per feste è guico.' contains paragraphs devoted to 'Rochette', 'Girandola semplice', 'Scuto artificiato', 'Tromba da guico artificiata', 'Girandola doppia', 'Tore artificiata', and 'Ruota artificiata'.

References: Cockle 673. Brock (1598)

CHAPUIS, N

Considerations sur l'Art des Feux d'Artifice en France ... et sur la necessité d'établir une salle d'artifice dependante des Menus-Plaisirs du roi par N. Chapuis ...

C020.1
L'Auteur: Paris 1830
20p. 8vo. NC — BL —

BROCK's bibliography in the 1949 edition spells it CHAPNIS and changes the country from 'France' to 'Autriche'. However an examination of his amendments and additions to his own copy of the 1922 edition (in this author's collection) clearly shows that these are errors of transcription.

References: BNC

CHERTIER, F. M

Essai sur les Compositions qui donnent le plus belles Couleurs dans les Feux d'Artifice. [*Essay on the compositions which give the most beautiful colours in fireworks.*]

C030.1
Delaunay: Paris 1836
88p.(inc.Tabs.). 20cms. NC 0344959 BL —

Nouvelles Recherches sur les Feux d'Artifice ... [*New research in fireworks.*]

C030.2
Chez l'Auteur: Paris 1843.
½T+TP+ii+682illus. 22cms. NC 0344960 BL —

C030.3
Chez l'Auteur/Mallet-Bachelier: Paris 1854
½T+TP+iv+638illus. 22cms. NC — BL 8716.bbb.10

Fuegos Artificiales. Modo de Hacer Varios Colores para Estos. Contiene 24 Recetas Buenas y Esperimentadas ... par J.L. y A. [*Artificial fireworks. Method of making various colours.*]

C030.4
Imprenta y libreria de Cristobel y J.M. Magallon: Zaragoza 1853
½T+TP+42+iii. NC 0417023 BL —

Auszug des Neuen & Beachtenswerthen aus F.M. Chertier's Lustfeuerwerkerei. Nach dem Franzoischen mit Anmerkungen von Martin Websky.

(See M. WEBSKY's *Lustfeuerwerkkunst* 1846)

Der excellirende Lustfeuerwerker ...

(See under BUETTNER, F.C.A.)

Chertier is generally considered to be one of the great pioneers of present day pyrotechny. His *Essay* of 1836 is a landmark, as it introduces all the basic colours into fireworks for the first time. His 1854 edition introduces a number of other important chemicals into the pyrotechnist's range and many of his experiments in the mid-19th century have been 're-discovered' by later workers. As Michael Swisher says, "a profound intuitive chemist whose work is the foundation on which much subsequent work stands."

The Spanish edition of 1853 is a partial verbatim translation of the 1843 edition made by an author who had lived in France and seen the results of Chertier's experiments. See also under RUGGIERI.

References: Du Pont. CP (1843, 1854). BC. Brock (1854)

CHIKOLEV, V. N

Rukovodstvo k prigotovleniyu i szhiganiyu feyerverkov s opisaniyem ustroystva elektricheskgo osveshcheniya.

[Guide to the preparation and lighting of fireworks with a description of the electrical illumination principle.]

C040.1
Moscow 1868

NC	—	BL	—

References: SHIDLOVSKY's bibliography.

COCKLE, Maurice J. D

A Bibliography of Military Books up to 1642.

C050.1
Simpkin Marshall: London 1900

TP(BR)+xxxviii+268. 26cms.	NC 0508532	BL 2753.s.3	

C050.2
Holland Press: London 1957.(2nd ed)

TP+xxxviii+268. 26cms.	RLC 0109944	BL —

C050.3
Holland Press: London (1978) (Reprint)

Eng.TP+TP+xxxviii+268+xiii. 26cms.	NC —	BL —

Although primarily concerned with military books, it nevertheless provides a useful source of information concerning a number of the titles that appear in this Bibliography, since many early military books contain chapters on recreative fireworks. The first edition was limited to 250 copies.

References: CP (1978)

COLLECÇÃO DE RECEITAS

Additamento ao VI. volume da Collecção de Receitas e segredos particulares, necessarios para a maior parte dos artistas, contendo um pequeno tratado de Fogo D'Artificio. *[Additions to Volume VI of the Collection of Receipts and secrets, containing a short chapter on artificial fireworks.]*

C060.1
Segunda edição. Typographia de Cobellos: Lisboa 1850

TP+48+2Pls. 17cms.	NC —	BL —

References: BC

COLL(I)ADO, Luys (Louis, Luigi)

Pratica Manvale di Arteglieria nellaquale si tratta della inuentione di essa, dell'ordine di condurla, & piantarla sotto à qualunque fortezza, fabricar mine da far volar in alto le fortezze, spianar le montagne ... far fuochi artificiali, con altri bellissimi secreti all'essercitio dell'arteglieria appartenenti. [*Practical manual on artillery which deals with its invention, its use and positioning under any stronghold, levelling mountains ... making artificial fires, with other profound secrets in the use of artillery.*]

C070.1
P. Dusinelli: Venetia M D LXXXVI (1586)
TP+x+92L(inc.31illus.)+2Pls. 33cms. NC 0537349 BL C.54.k.2 +

Platica Manval de Artilleria, en la qval se tracta de la excelencia de el arte militar, y origen de ella, y de las maqvinas con qve los antigvos començaron a vsarla, de la invencion de la polvora, y artilleria, De el modo de conduzirla, y plantaria en qualquier empressa, Fabricar las Minas para bolar las Fortealezas, y Montan as, Fuegos artificiales, Varios secretos, al arte de la Artilleria ... muy necessarios.

C070.2
Pablo Gotardo Poncio: Milan 1592
TP+x+225(illus.)+5FPls. 33cms. NC 0537352 BL C.73.g.9

Prattica Manvale dell'Artiglieria ... si tratta dell'eccellenza, & origine dell'Arte Militare, e delle Machine vsate da gli antichi; dell'inuentione della Poluere, & Artiglieria; ... di Fuochi Artificiali; e di varij secreti ... Composta da Luigi Colliado ... [*Slight variants in long-title.*]

C070.3
G. Bordoni & P. Locarni: Milano 1606
TP+xviii+388(illus.)+16Pls. 27cms. NC 0537353 BL 534.i.7

C070.4
Filippo Ghisolfi & Gio B. Bidelli: Milano 1641
TP+xiv+328(inc.57figs.)+27Pls. 25cms. NC 0537355 BL 15O2/213 +

Collado wrote the first edition in Italian in 1586. The second edition of 1592 he wrote in his native Spanish and this was subsequently translated into Italian again to give the editions of 1606 and 1641. There were only a few additional paragraphs added to the 1586 edition but a large number of extra plates. The Italian editions of 1606 and 1641 are virtually identical although the dedicatees are different, the text has been reset and the layout of the plates altered; many of the text figures in the 1606 edition becoming separate plates in the 1641 version. In the first two versions only the leaves have been numbered, in the last two the pages have been numbered. The larger page size of the Spanish edition means that many of the illustrations have been printed in the text which accounts for the larger number of inserted plates in the two Italian editions.

The great interest of this work so far as pyrotechnists are concerned is contained in Chapter 11 which is entitled, 'Of ... fires which can be adopted for use in times of festivity'. Although the 1586 edition does not contain any plates the editions of 1592, 1606 and 1641 include three figures depicting recreative fireworks, a line rocket, a vertical and horizontal wheel, and a device (called in Spanish 'Las Granadas') which consists of a thick walled tube, filled with a slow burning composition, and drilled with many angled holes each of which contains what looks like a small rocket. The text of the 1586 edition describes these items, but they are not illustrated. This book contains some of the earliest and fullest descriptions of recreative fireworks.

References: Cockle 664 & 668. E.T.y Guell *Bibligrafia Espanyola D'Italia* (Barcelona 1927). CP (1641)

COMPLEAT, Gunner (The) See VENN, Thomas

COMPLETE

The Complete Art of Making Fireworks: improved to the Modern Practice; from the Minutest to the Highest Branches; containing Copious Directions for making Squibs, Crackers, Serpents, Rockets, &c. &c. ... by the Juvenile Pyrotechnician.

C080.1
Printed and Published by B. Hodgson: London c.1830
PC+ColFP+TP+26(-4). 18cms. NC — BL —

C080.2
Orlando Hodgson: London c.1835
PC+ColFP+TP+24(-2). 19cms. NC — BL 1507/374

The Complete Conjuror; with the whole Art of Making Fireworks. By the Wizard of the North.

C090.1
J. Catnach, 2 & 3 Monmouth Court, 7 Dials: [London] c.1833
ii+3-12p. 17cms. NC — BL —

The Complete Conjuror. Containing Various Wonderful Tricks, with Cards and Dice also the Art of Making Fireworks.

C100.1
J. Russell: Birmingham c.1840
TP+12p. 13cms. NC — BL —

The Complete Art of Making Fireworks appears to be a very similar volume to that listed under *The ART of Making Fireworks*. It appeared in a Sotheby sale in December 1982 of the Findlay collection and Magic Circle books.

References: Stott 174/5. Hall 65. CP (O.Hodgson)

COURS

Cours abrégé d'artifices, contenant la confection, la réception, la conservation et la démolition des munitions et artifices de guerre; suivi de notions sur les artifices de joie. [*A short course of fireworks, containing the making, the acceptance, the storage and the demolition of munitions and munitions of war, followed by ideas on recreative fireworks. (2 vols. Vol.I Text, Vol.II Plates.)*]

C110.1
Ministre de la Guerre: Paris and Strasburg 1850
(I) ½T+TP+iv+438+1F.Tab. 21cms
(II) TP+20Pls.+1FT. 21cms. x 28 cms. NC 0746694 BL 1262.d.6/7

The BL copy of this intriguingly titled volume is, regrettably, missing; however I was able to examine a copy supplied by the Landesbibliotek of Strasburg. This rare, curious and unusual book consists of a summary of the work carried out by the 'School of Pyrotechnics' at Douai, Metz, Strasburg and Toulouse during the latter half of the 1840s. The first three-quarters deals only with military pyrotechnics, but the last eighty or so pages are devoted to the manufacture of 'Artifices de Joie'. It is truly a 'Manufacturers' bible as it describes in detail each of the operations needed to make a wide range of pyrotechnic items. The formulas given show that already mixtures of sulphur and potassium chlorate were in general use for the production of colours. There are detailed lists showing, for every item, the number of workers required, and the precise type, size and quantity of all the tools and materials necessary, the way the work should be organised, and the number of items that could be produced in a given period. At the end is a very comprehensive table listing all the items used in a large display with an analysis of the cost of each, both as regards labour and materials. The book of Plates, all of which are double, contain five that relate to recreational fireworks. Four of these describe set-pieces and lancework items, including a Salamander and a triple rotating pyramid as well as some lance-work. The final table lists in great detail the dimensions, contents and quantity of all the rockets, candles, shells, jets, gerbes and lances that were used for a display given at Metz on April 3rd 1843.

References: Lorenz

CUTBUSH, Professor James (1788-1823)

A System of Pyrotechny, Comprehending the Theory and Practice, with the Application of Chemistry; Designed for Exhibition and for War. In Four Parts ...

nc2
JN5 JN10
SG 1379/$110

C120.1
Clara F. Cutbush: Philadelphia 1825
TP+xlii+612+1Pl+ii. 23cms. NC 0851536 BL —

J.B. Barkley in *Pyrotechnica No.5* (Oct.1979) says:- 'This voluminous treasure discusses every aspect of Pyrotechny, both cilvilian and military, known to the author at the time. Cutbush obviously consulted every contemporary book and journal, domestic and foreign, dealing with pyrotechny to prepare such an exhaustive and scholarly work; every little detail was set down. The book is a landmark, not only in the general history of pyrotechny, but especially as the first major American contribution to the literature of Fireworks.' It was published posthumously by his wife.

References: Brock. Du Pont. CP. Rink 2158

D * * *, J. E

Anweisung zur Lust- Ernst- und Wasserfeuerwerkerey für Artilleristen und Freunde dieser wissenschaft. [*Instructions for recreative, military and water fireworks for artillerymen and lovers of this science.*]

D010.1
W.G. Kircher: Goslar 1802
TP+iv+112 19cms ND — BL —

References: Barfod

DAMBACH, Christoff

Büchsenmeisterey ... sampt getrewlichen vnderweisung mancherley künstlicher Fewrwerck: Als nemlich der Petardt Spreng vnd andern Kugeln zu Schimff vnd Ernst zu Wasser vnd Land zu gebrauchen nützlich so vor niemals in Truck aussgangen ... [*Gunnery ... with reliable instructions in many kinds of artificial fireworks, such as petards and other bombs for recreative and military purposes to be used on water and land never printed before.*]

D020.1
W. Hoffmanns: Frankfort-am-Mayn 1608
TP(BR)+vi+164illus. 20cms. ND — BL —

D020.2
Conrad Corthoys: Franckfurt am Mayn 1615
TP(BR)+ii+FP+206+Eng. Colophon. 19cms. ND — BL —

Describes line-rockets, wheels, grenados and simple machines. The 1615 edition contains an extra section on siege and counter-siege.

References: RDML. Fahler (1608). Lotz (1609). Jähns.

DANILOV, M. V

Dovol'noye i yasnoye pokazaniye, po kotoromu vsyakoy sam soboyu mozhet prigotovlyat' i delat vsyakiye feyerverki i raznyya illyuminatsii. [*Sufficient and practical instructions by which anyone can prepare and make fireworks and various illumination.*]

D030.1
Universitetskaya tipografiga: Moscow 1779
ND — BL —

References: SHIDLOVSKY's bibliography.

DAVIS, Tenney Lombard (1890-1949)

The chemistry of Powder and Explosives.

D040.1
J. Wiley: New York/Chapman Hall: London. 1941/43
TP+viii+490illus. +viii 21cms. ND 0082218 BL —

D040.2
(Reprint)Angriff Press: Hollywood [1970]
TP+viii+490illus.+viii. 21cms. ND 0082219 BL —

As Davis states in his introduction, the 'book is primarily intended for chemists' and for 'the use of students in the course in powder and explosives ... at the Massachusetts Institute of Technology', where the author taught for 25 years. Much is therefore of a technical and chemical nature. Two chapters however are devoted to black powder and civilian pyrotechnics, including a detailed description of the manufacture of Chinese firecrackers. The author was for some time technical director of the National Firework Co. of Boston. A full review appeared in the July 1972 issue of the magazine, *American Pyrotechnist Fireworks News*.

References: Brock (1943). Du Pont. CP (1970)

DEMIDOV, A. P

O proiskhozhdenii uveselitel'nykh ogney, izobretenii porokha i sistematicheskoye opisaniye raketnykh pavilionov. [*Origins of illuminating fires, invention of powder and a systematic description of rocket pavilions.*]

D050.1
St. Petersburg 1820
ND — BL —

References: SHIDLOVSKY's bibliography.

DENISSE, Amédée

Traité Pratique Complet des Feux d'Artifice. [*Complete practical treatise of fireworks.*]

D060.1
Chez L'Auteur: Paris 1882
½T+TP+Pl+vi+Pl+428(inc.1Pl+illus.). 22cms.
ND 0168710 BL 8716.bbb.29

Les progrès de la Pyrotechnie.

D060.2
Paris 1885
PC+TP+8. 22cms. ND — BL —

Feux d'artifice, formules nouvelles, fusées volantes moulées, notes diverses relatives a la confection des pieces pyrotechniques. [*Fireworks, new formulae, rocket moulds, miscellaneous notes relating to pyrotechnic items.*]

D060.3
Poulenc frères: Paris 1886
PC+TP+73(inc.Pls.). 22cms. ND — BL —

Feux d'artifice sifflants ... [*Firework whistles.*]

D060.4
Poulenc frères: Paris 1888
20+Pl+figs. ND — BL —

The two Plates, in *Traité Pratique,* are engravings of Chertier and Tessier. The author was, apart from being a prolific and innovative pyrotechnist, a commercial artist and, as a result, the numerous excellent illustrations are particularly clear, detailed and elaborate. His books go into great practical detail and extract the best from CHERTIER and TESSIER. The *Progrès* published in 1885 is an account of 'l'affaire Poussier de tonneau' which concerned a Judgement exempting pyrotechnic manufacturers from the State's monopoly of blackpowder production. This report is reprinted again at the beginning of his 1886 work *Fusées volantes ...* which contains many additional formulaes as well as improvements and additions to the main work, *Traité Pratique*. The work on whistle compositions of 1888 contains the first detailed examination of this curious and entertaining effect in the pyrotechnic literature.

References: Brock (1882). BNC(All). CP (1882, 1885, 1886)

DESORTIAUX, E

Traité sur la Poudre, les Corps Explosifs et La Pyrotechnie, par les Docteurs J. Upmann & E. von Meyer. Ouvrage traduit de l'Allemand revu et considérablement augmenté par E. Désortiaux. [*Treatise on gunpowder, high explosives and pyrotechnics.*]

D070.1
Ch. Dunod: Paris (1878)
½TP+TP+xxxiv+860+8FPls. 25cms. NU 0288539 BL 8903.h.10

The translation by Desortiaux was made from *Das Schiesspulver* by J. UPMANN and *Die Explosivkorper und die Feuerwerkerei* by E. von MEYER. Deals in the main with the manufacturing techniques in Europe for the commercial production of gunpowder and high explosives, together with amorces, pyrotechnic fuses and coloured fires.

References: Brock (1882[?]). CP (1878)

DIETRICH, Major A. L

Die Kunst-Feuerwerkerei; oder, Gründliche anweisung zum anfertigen, aufstellen und abbrennen von Land und Wasserfeuerwerk nach den neuesten Erfahrungen zusammen gestellt durch 101 Abbildungen ... [*The art of making fireworks; or basic instructions for making, setting up and firing land and water fireworks according to the latest developments, with 101 illustrations.*]

D080.1
Potsdam 1845
ND — BL D

D080.2
F. Rubach: Leipzig c.1845
viii+232+11FPl. 20cms. ND 0264955 BL —

DILETTANTE

L'Arte di Fare I Fuochi d'Artifizio con poca spesa per le feste di Famiglia, sponsali ed altre simili occasioni di un Dilettante. [*The art of making fireworks at little cost for family celebrations and other similar occasions for amateurs.*]

D090.1
(First edition). Giambattista Sonzogno: Milano 1819.
PC+½T+TP+88(-4)+10Pls.(1F). 17cms. ND — BL —

D090.2
Seconda Edizione. Fratelli Sonzogno: Milano 1823
PC+ii+TP+88(-4)+10Pls.(lF). 17cms. ND — BL —

D090.3
Terza Edizione. Sonzogno: Milano 1833
½T+TP+88(-4)+10Pls.(1F). 17cms. ND — BL 1607/552

D090.4
Terza Edizione. : Napoli 1834
ND — BL —

D090.5
Quarta Edizione. Gaetano: Napoli 1847
PC+TP+54(-4)+vi+10Pls.(1F). 18cms. ND — BL —

These works all bear a strong family resemblance to the book first published in 1818 by AUDOT *L'Art de Faire a peu de Frais les Feux d'Artifice*. The Milan editions were all published by Cesare SONZOGNO who later became the editor of *Il Pirotecnico Moderno*. The only ones that I have been able to examine are identical except that the type has been reset.

References: CP (1819, 1823, 1847)

DMITRIYEV, G. Kh

Podarok dlya lyubiteley feyerverkov. [*Offering to firework lovers.*]

D100.1
1880
ND — BL —

References: SHIDLOVSKY's bibliography.

DUSSAUCE, Hippolyte (1829-1869)

A Practical Treatise on the Fabrication of Matches, Guncotton, Coloured Fire and Fulminating Powders.

D110.1
H.C. Baird: Philidelphia 1864
xv+336illus. 19cms. ND 0459286 BL 7943.bbb.36

References: Brock. Du Pont

EGUCHI, Harutaro (1924-)

Hanabi monogatari. [*The History of Fireworks.*]

E010.1
Chunichi shimbun-sha: Nagoya 1982
229+10photos 18 cms. NE — BL —

SHIMIZU describes this as, 'a beautiful book for every one'. From the list of contents it is clearly an important and detailed history of fireworks in Japan. Many

17th, 18th and 19th century firework festivals and manufacturers are described as well as painters and writers who have dealt with the subject of fireworks.

References: CP

EISMA, A

De kunst om op eene gemakkelijke wijze allerlei soorten van vuurwerken te maken. Naar de derde Fransche uitgave, vertaald door A. Eisma. [*The art of easily making all sorts of fireworks. According to the third French edition, translated by A. Eisma*]

E020.1
L. Schierbeek: Leeuwarden 1839
77+10Pls. 20cms. NE — BL —

Not listed by Brinkman. Probably a translation of AUDOT's 3rd edition of 1825.

References: Du Pont

ELLERN, Dr. Herbert

Modern Pyrotechnics. Fundamentals of Applied Physical Pyrochemistry.

E030.1
Chemical Publishing Co.: New York 1961
TP+320(-2). 22cms. RLE 020372 BL —

Military and Civilian Pyrotechnics.

E030.2
Chemical Publishing Co.: New York 1968
Photo+TP+viii+464. 22cms. NE — BL —

The first edition, as the introduction states, 'attempts to give a broader view of exothermic dry reaction chemistry and physics ... for the practical worker in the field.' Althogh mainly concerned with military applications, it includes a number of formulae. The second edition greatly increases the scope and number of formulae although their application to recreative purposes is limited. It includes a brief chapter on fireworks by LANCASTER.

References: CP (1961, 1968)

ELRICH, Daniel See SIMIENOWICZ

[ELRINGTON, J. F]

Practical hints on Pyrotechny and the Art of Making Balloons for the use of Amateurs.

E040.1
W. Underwood: Dublin 1850
Cover+TP+iv+108+6Pls. 19cms. NE — BL PO6042

The cover-title is 'The Art of Making Fireworks'. The title-page is inscribed in the Patent Office Library copy as, 'from the author J.F. Elrington'; his name does not otherwise appear. This book is mentioned by W.H. Browne in the introduction to his *The Art of Pyrotechny* as being, 'the simplest and best that has been published in recent years'. It is also listed by Hanbury in his private manuscripts that I possess.

ENDLESS AMUSEMENT

Endless Amusement; a collection of upwards of 400 Entertaining and Astonishing Experiments. Among a Variety of other Subjects, are Amusements in ... The Art of Making Fire-Works ... [*There are small variants in the long-title.*]

E050.1
Printed by Gye and Balne. Published and Sold by Thorp and Burch: London c.1818
IFPl+TP+x+216(-12). 16cms. NE — BL 7923.d.44

Endless Amusement ... To Which is Added, A Complete System of Pyrotechny or the Art of Making Fireworks ...

ENGLISH EDITIONS

E050.2
2nd ed. For Thorp and Burch/Thomas Boys: London c.1819
1FPl+TP+x+216(-12). 15cms. NE 0120832 BL 8705.aa.4

E050.3
3rd ed. For Thorp and Burch/Thomas Boys: London c.1820
TP+x+1FPl+216(-12). 15cms. NE — BL 1035.e.28/1

E050.4
4th ed. For Thorp and Burch/Thomas Boys: London c.1825
1FPl+TP+x+216(-12). 15cms. NE 0120835 BL —

E050.5
Fifth Edition. J. Gilbert: London (c.1830)
1FPl+TP+x+216(-12). 15cms. NE 0120836 BL —

E050.6
Sixth Edition. H. Washbourne & T. Tegg: London MDCCCXXXIV (1834)
FPl+TP+x+216(-12). 15cms. NE — BL —

E050.7
Seventh Edition. H. Washbourne & T. Tegg: London MDCCCXXXIX (1839)
FPl+TP+x+216(-12). 15cms. NE 0120837 BL —

AMERICAN EDITIONS

(First Edition?)

E050.8
Second Edition. M. Carey: Philadelphia 1821
216p (Plates and woodcuts) . NE 0120833 BL —

E050.9
Third Edition. M. Carey: Philadelphia 1822.
216p. 16cms. NE 0120834 BL —

E050.10
W.C. Borradaile: New York 1831
xii+216(-12). 15cms. NE — BL —

E050.11
From 7th London Edition. Lea & Blanchard: Philadelphia 1847
xii+218p. 17cms. NE 0120838 BL —

E050.12
Clarke, Austin & Smith: New York 1853
201p illus. 18cms. NE 0120839 BL —

In the First Edition *The Art of Making Fireworks* occupies some 15 pages and deals with 'touch-paper', 'crackers', 'squibs and serpents', 'gunpowder', and various types of rocket, vertical and horizontal wheels and fires for the same. Also the use of cast iron for making 'Chinese flowers'. By the 4th edition this section has been expanded to some 40 pages and includes several rather sophisticated fireworks including shells and water-fireworks. I have not been able to locate any reference to a 'First American edition'. Even Rink's *Technical Americana* does not mention it. If so, then the 'Second Edition' of 1821 shares, with GROTZ's *The Art of Making*

Fireworks, the privilege of being one of the first books published in America to deal with the subject of Fireworks. Nevertheless, CUTBUSH's classic work *A System of Pyrotechny* (1825) is still the first true American work on the subject. There is a companion volume called *A Sequel to the Endless Amusements* but this does not contain any material dealing with Fireworks.

References: Hall 116-123. Stott 255-267. CP (1825). Rink 192-193

ENGLE, M.B

The Manufacture of Fireworks with Home-made equipment.

E060.1
National School of Chemistry: California 1956
NE — BL —

Michael Swisher comments that this book is something of an oddity since "it is a mixture of plagiary and misinformation, much of which appears to be a translation from (of all sources) MOREL!". Also there is a section on whistles translated from DENISSE (1888) with plagiarized formulae.

ESCHENBACHER, August

Die Feuerwerkerei; oder, die Fabrikation der Feuerwerkskörper. [*The making of fireworks.*]

E070.1
Hartleben: Wien, Pest, Leipzig 1876
TP+iv+287+iiads.(inc.26 figs). 19cms. NE 0163420 BL PO19437

E070.2
2e verm. u. verb. aufl. Hartleben: Wien, Pest, Leipzig 1885
iv+327(inc.49figs.). NE — BL —

E070.3
3e verm. u. verb. aufl. Hartleben: Wien, Pest, Leipzig 1897
TP+vi+272(inc.51figs.). 19cms. NE 0163421 BL PO5778

E070.4
4e verm. u. verb. aufl. Hartleben: Wien 1925
viii+216(inc.52figs.). NE 0163422 BL —

References: Kayser. E&O (1885, 1897)

ETTEN, Henri van

Recreation Mathematique ...

E080.1
Jean Appier Hanzelet: Pont-à-Mousson 1624

E080.2
Jean Appier Hanzelet: Pont-à-Mousson 1626

E080.3
(2nd ed.) R. Boutonne: Paris 1626

E080.4
Anthoine Robinot: Paris 1626

E080.5
(3rd ed.) Jean Moreau: Paris 1626

E080.6
(4th ed.) R. Boutonne: Paris 1627. (Augm. par DHPEM)

E080.7
C. Rigaud & C. Obert: Lyon 1627

Recreations Mathematiqves. Composées de plusieurs Problemes, plaisans & facetieux, d'Arithmetique, Geometrie, Astrologie, Optique, Perspective, Mechanique, Chymie, & d'autres rares & curieux Secrets: Plusieurs desquels n'ont jamais esté imprimez. Premiere et Seconde partie. La troisieme Partie, contient vn Recueil de plusieurs gentilles & recreatiues inventions de feux d'artifice: La maniere de faire toutes sortes de fuzées, simples & composées: le tout represente par figures. [*Mathematical recreations. Containing many problems, pleasant and humorous ... the third part containing a collection pleasing and amusing inventions of fireworks; the manner of making all sorts of rockets simple and compound, all represented by figures. (Small variants in long-title.)*]

E080.8
Charles Osmont: Roven M.DC.XXVIII. (1628)
TP+ii+22O+ii+TP+iv+88+TP+ii+48+ii+78WC. 16cms.
NL — BL —

E080.9
Charles Osmont: Roven 1629. NL 0303616 BL 529.b.14

E080.10
Pont-à-Mousson 1629. NL — BL —

E080.11 Charles Osmont: Roven 1630.	NL	0303618	BL	—
E080.12 Charles Osmont: Roven 1634.	NL	—	BL	—
E080.13 Paris 1638.	NL	—	BL	—
E080.14 Paris 1639.	NL	—	BL	—
E080.15 Claude Prost: Lyon 1642.	NL	0303622	BL	—
E080.16 Claude Prost: Lyon 1653.	NL	0303624	BL	—
E080.17 J.B. de Ville: Lyon (1656).	NL	—	BL	—
E080.18 Cardin Besongne: Paris 1660.	NL	—	BL	—
E080.19 Paris 1661.	NL	—	BL	—
E080.20 Claude Prost: Lyon 1669.	NL	0303625	BL	—
E080.21 J.B. de Ville: Lyon 1680.	NL	0303626	BL	—

(ENGLISH editions. Translation erroneously attributed to OUGHTHRED.)

Mathematicall Recreations. Or a Collection of sundrie Problemes, extracted out of the Ancient and Moderne Philosophers, as secrets in nature, and experiments in ... Fireworks ... Most of which were written first in Greeke and Latine, lately compiled in French, by Henry Van Etten Gent. And now delivered in the English tongue, with the Examinations, Corrections, and Augmentations.

E090.1
T. Cotes for Richard Hawkins: London 1633
ii+EngTP+TP+xxxvi+286+iv. 17cms. NL 0303596 BL 52.a.2

Mathematicall Recreations ... Wherunto is added the Description and Use of the General Horological Ring: and The Double

Horizontal Diall. Invented and written by William Oughtred. [*Added TP 1652*]

E090.2
William Leake: London MDCLIII (1653)
ii+EngTP+TP+xxxvi+286. 16cms. NL 0303600 BL 529.b.15

E090.3
William Leake and John Leake: London 1674
EngTP+TP+xxxvi+282. 17cms. NL 0303607 BL 8530.aaa.30

It seems likely from Trevor HALL's researches that *Recreation Mathematique* was probably written, or perhaps 'compiled' would be a better word, by LEURECHON's pupil van ETTEN. Both LEURECHON and ETTEN were real people and not pseudonyms as is sometimes thought. However these editions are generally catalogued under 'LEURECHON' in the NUC - hence the prefix 'L'. In many ways the Rouen edition of 1628 is by far the most important since it was the first to contain a section on artificial fireworks and as a result all subsequent works by such as WHITE, OZANAM, MYDORGE, HUTTON and others obviously felt that Fireworks should be included in their own compilations. It is also the first edition in which the main title becomes plural *Recreations Mathematiques*. As Trevor HALL suggests in his essay on the subject the firework section was probably supplied by Jean APPIER. The work aroused a great deal of interest as can be seen from the number of commentary works that it spawned, by HENRION and MYDORGE, and the number of translations and editions that were subsequently published. Even Backer is confusing on the subject, 'Il existe plusieurs autres editions de ces Recreations, soit avec les notes (by MYDORGE and HENRION), soit sans les notes.'

The first English edition of 1633 was, suggests Mr. HALL, made by Thomas MALTHUS and not by William OUGHTRED. Translations were also made into German, (Daniel SCHWENTER), and Dutch, (Wynant van WESTEN).

Sotheran has lengthy notes on some of these editions. In reference to the 1628 and 1669 editions he says, 'Neither ... was known to Brunet or Grasse who mention however others of Rouen 1629 and Lyon 1680.' Of the English editions he says, ' A highly interesting storehouse (once of great popularity, and now rare in all its editions) of a multitude of mathematical, astronomical and physical problems. It was based on the work of Bachet de Meziriac and included, beside numerous original problems by the author, some taken from Cardan; and in its turn formed the foundations of the works of Mydorge, Ozanam, Montucla & Hutton.' Of the third and last edition of 1674 (SOT10774: 1921 ed.) he says, ' ... the rarest of all. Being unknown to Lowndes, Watt and Allibone.' Full details of most of these editions can be found in Backer. See also the note in the Introduction.

References: Backer (1624, 1626, 1626, 1627, 1629, 1629, 1638, 1639, 1642, 1660, 1661, 1680). BNC (1628, 1630, 1634, 1642). Eng. Eds. STC10558.5. Du Pont 1634. Hall 282-4. Stott 429-31. CP (1628, 1633). STC10558.5 (1633). Wing 1790 (1653), 1791 (1674)

EVIEUX, Eugene A (1883- ?)

Manuel de l'Artificier; artifices, poudres, explosifs ... [*Firework manual; fireworks, gunpowder, explosives.*]

E100.1
J.B. Bailliere et fils: Paris 1923
396illus. 16cms. NE — BL —

References: Du Pont

FABER, Henry Burnell (1877-1938)

Military Pyrotechnics. (Vol.I - 'History'. Vol.II - 'Manufacture'. Vol.III - 'Chemicals'.)

GW 11/49/£40
SG 1253/303/$130

F010.1
Government P.O.: Washington 1919
(Vol.I) TP+Pl+230(-4). (Vol.II) TP+288(-2).
(Vol.III) TP+230(-2). 23cms. NF 0002223 BL A.5677/8

This curious hotch-potch of a work contains amongst other wayward pieces of information, a literal translation of part of PERRINET d'Orval's *Manuel de L'Artificier* (1757) including the plates.

References: Du Pont. Brock. CP

FAHLER, Eberhard

Feuerwerke des Barock. Studien zum öffentlichen Fest und seiner literarischen Deutung vom 16. bis 18 Jahrhundert. [*Baroque fireworks. Studies concerning public celebrations and its literature from the 16th to the 18th century.*]

F020.1
J.B. Metzlersche: Stuttgart (1974)
½T.+TP+ii+248illus. 24cms. NF — BL —

References: CP

FALBESONER, Hartmann

Lustfeuerwerkerei für Berufsfeuerwerker und Liebhaber. Grundliche Anweisung zur Herstellung aller gegenwartig gebrauchlichen Feuerwerkskorper und deren Zusummenstellung zu Feuerwerken ... [*The making of fireworks for fireworkers and amateurs. Basic instruction for the manufacture of all currently used firework items and their assembly into fireworks.*]

F030.1
A. Hartelben: Wien und Leipzig 1909
vii+236(100illus.)+ivads. 23cms. NF0018850 BL —

References: Du Pont. Kayser. G&O

FAVE, General Ildephonse See REINAUD, J. T

FEDOROV, P. A

Pirotekhnik-lyubitel'. [*Amateur pyrotechnist.*]

F040.1
1904
NF — BL —

References: SHIDLOVSKY's biliography.

FERNANDEZ DE MEDRANO, Sebatian

El Perfecto Bombadero, y Practico Artificial. Dividido en dos tratados, que el Primero contiene los Artificios de fuego de regocijo, y el Segundo, los Marciales, nueva invencion de Morteros, Bombas, Carcasces, Granadas, que saca à luz ... [*The perfect bombardier, and practical fireworks. Divided into two parts, the first containing fireworks for rejoicing, and the second for war, new inventions of mortars, bombs, shells, grenades.*]

F050.1
Francisco Foppens: Brusselas 1691
TP+xiv+118+3FPLs. 15cms. NF 0093167 BL 8823.aa.36

The first part deals in some detail with the preparation of saltpetre, sulphur and gunpowder and the manufacture of rockets, wheels, petards and letters of fire. The second part deals only with military matters. He wrote a number of other books with similar titles but they are only concerned with warfare.

References: Palau y Dulcet 89221

FINSTERWALD, Caesare

Nuovo trattato di Pirotecnia, ovvero l'Arte dei fuochi d'artifizio. Opera originale contenente tutti i piu recenti ritrovati dell'arte pirotecnia. [*New treatise of pyrotechnics, or the art of fireworks. An original work containing all the latest discoveries in the art of pyrotechnics.*]

F060.1
Amati: Trieste 1889
86p. 8mo. NF — BL —

References: Pagliaini. G&O

FIREWORKS

Fireworks and How to Make them.

F070.1
G. Routledge: London (1871).
ColPC+Adsiii+½T+TP+ii+63(-6)+ivads. 17cms.
NF — BL 8715.aa.19

F070.2
G. Routledge: London c.1885
iiiads.+½.001TP+ii+63(-6)+iiiads. 17cms. NF 0151153 BL —

F070.3
G. Routledge: London c.1887
iiiads.+½T+TP+ii+63(-6)+iiiads. 17cms. NF — BL —

This little practical work was 'issued' four times between 1871 [First Issue?] and 1887, the 'Fourth Issue'. The text was the same but the advertisements, some of which are dated, changed.

References: Du Pont. CP c.1885 & 1887

FLOBECK, Zach

Praktisk Lårobok i Fyrverkerikonsten. [*Practical manual of the art of fireworks.*]

F080.1
N.M. Lindh: Orebro 1823
TP+ii+76+2FPls. 18cms. NF — BL —

References: Barfod

FRANCESCO, T. de

La Moderna pirotecnica per tutti. Manuale practico per la Fabbricazione dei Fuochi Artificiali. [*Modern pyrotechnics for everyone. A practical manual for the manufacture of fireworks.*]

F090.1
G. Lavagnolo: Torino c.1950
PC+TP+ii+121(-4)+ii+ivads(56figs). 24cms. NF — BL —
Contains detailed descriptions of the making of various types of cylinder shells as well as flash powders, aluminium streamers and tremalon stars.

References: CP

FREY, Oscar

Die Feuerwerkskunst. ... insbesondere für Liebhaber und Anfanger bearb. [*The art of fireworks, especially adapted for amateurs and beginners.*]

F100.1
Bartholomaus: Erfurt 1877
xi+135(101illus.). NF — BL —

F100.2
Zweite verhessrte Auflage. Fr. Bartholomaus: Erfurt [1885]
Col.Eng.Cover+TP+viii+154+16FPls. 24cms. NF — BL PO22929

The engraving on the cover of the [1885] edition is a lurid one of demons dancing round a fire with a few fireworks in the background.

References: Brock. Kayser. G&O (1885)

FREZIER, Amedée Francois (1682-1773)

TRAITÉ/ DES FEUX D'ARTIFICE / POUR LE SPECTACLE / ou l'on Voit / ... / Par le Sieur F ... [*A treatise of fireworks for celebration.*]

F110.1
A PARIS/ Chez DANIEL JOLLET ... / M.D.CC.VI. (1706)
xxiv+349+x+8Pls(in Text). 16cms. NSF0045421 BL 1568/2239

TRAITÉ/ DES / FEUX D'ARTIFICE / ou l'on voit / ... / Par le Sieur F ...

F110.2
A PARIS / Chez PIERRE FRANCOIS GIFFART .. / M.D.CC.XV (1715)
xxiv+394+x+8Pls(in Text). 16cms. NF 0380588 BL 55.a.3

TRAITÉ/ DES FEUX / D'ARTIFICE / ou L'ON VOIT / ... / Par le SR. FREZIER.

F110.3
A LA HAYE / Chez J.NEAULME Libraire / M.DCC.XLI. (1741)
EngTP+TP(BR)+xviii+394inc.8Pl+vi. 16cms. NF 0380589 BL —

TRAITÉ/ DES / FEUX D'ARTIFICE / POUR LE SPECTACLE / NOUVELLE EDITION, / Toute changée, & considerablement augmentée, / Par M. F * * * D.D.F.D.B. / ...

F110.4
A PARIS,QUAY DES AUGUSTINS, / Chez CH. ANT. JOMBERT, ... / M.DCC.XLVII (1747)
½T+EngTP+TP+lii+496+13FPls. 16cms. NF 0380590 BL 1400.k.10

F110.5
A PARIS, QUAY DES AUGUSTINS, / Chez NYON Fils, a l'occasion / ... M.DCC.XLVII (1747)
½T+EngTP+TP+lii+496+13FPls. 16cms. NF 0380591 BL 1560/1063+

F110.6
A PARIS / Chez CHARLES-ANTOINE JOMBERT / ... M.DCC.XLVII (1747)
½T+EngTP+TP+l+496+13FPLs. 16cms. NF — BL 8517.cc.18

Die kleine Feuerwerke: aus der Franzoische.

F110.7
Joachim: Leipzig 1803.
NF — BL —

Frezier was probably one of the most influential writers on pyrotechnics in the mid-18th century and some of his editions are still fairly common today. He was a military engineer by training but had the amateur's fondness for fireworks. His first book was written when he was only 24. He confesses that his enthusiasm for fireworks began when he saw the display in 1697 for the Treaty of Ryswick. The text and illustrations however are largely derivative of previous works, particularly SIMIENOWICZ, but purged of what Frezier rightly calls that author's, "fades et inutiles reflexions". Unlike many other authors he scrupulously cites his sources, such as Hanzelet [see APPIER], Henrion [see MYDORGE] and FRONSPERGER. The books contain much information about the conduct and planning of displays rather than detailed practical instruction for the manufacture of the individual items. As Michael Swisher remarks, "here, as so often, we find the phenomenon of a writer on pyrotechny who is not a practising pyrotechnist, but a 'gentleman amateur'; doubtless in Frezier's time, the actual firework makers were illiterate mechanics, and the men who ran the factories not about to publish their secrets."

Due to the considerable confusion that often exists concerning the various editions of Frezier's works and those of PERRINET d'Orval, full title details have been given for the 'first edition' of 1707 and its reissue of 1715, the 'pirated' edition of 1741 and the 'second edition' of 1747. The engraved title-page of the first edition dated 1706 has a date of 1707. The 'Approbation' is dated September 3rd 1705. The reissue by Giffart in 1715 is, apart from minor differences on the main title-page, identical to that of 1706. In 1741 Neaulme issued a pirated edition which again, apart from variations on the title-page, is identical to the 1706 edition. Encouraged by this pirated edition Frezier wrote a second edition which appeared in 1747 in three differing versions, two by Jombert and one by Nyon. The version issued by Nyon is identical to that issued by CH.ANT.JOMBERT apart from the title-page. The version issued by CHARLES-ANTOINE JOMBERT has been reset which gives differing catch-words to the 'DICOURS' (sic) and also differing page vignettes. See also note under PERRINET d'Orval.

References: Du Pont. Brock. Berlin 3300 (1715/47). Kayser(1803). CP (1741, 1747)

FRONSPERGER, Leonhard (Baron von Mindelheim). (? -1575)

Vonn Geschütz vnnd Fewrwerck, wie dasselb zuwerffen vnd schlissen Auch von gründlicher Zuberaitung allerley gezeugs vnd rechtem gebrauch der Fewrwerck wie man die in Schimff oder Ernst von der hand aus Fewrbüchsen oder Böllern zu lust oder gegen den Feinden sol werffen schiessen oder in stürmen in vnd aus den Besatzungen zubrauchen. [*Of artillery and fireworks. About the thorough preparation of all manner of equipment also how they are to be shot or launched from the hand, from firearms or mortars, for amusement, or against an enemy in attacks or defence of fortified places.*]

F120.1
David Zephel: Franckfurt am Mayn M.D.LVII (1557)
TP+x+58leaves. 31cms. NF — BL 1605/207+

F120.2
Johann Lechler: Franckfurt am Mayn 1564
TP+x+LIIIleaves+ii. 31cms. NF 0406913 BL C.66.h.11/4

This work originally formed part of his 'Kriegs ordnung'. In 1566 it was recast to form 'Book eight' of his *Von Kayerlichem Kriegsrechten Malefitz und Schuldhändlen Ordnung und Regiment* and published by Georg Raber in Franckfurt. Subsequent editions listed by COCKLE are 1571, 1578, 1598 and a Berlin reprint in 1819.

References: Du Pont (1564). Fahler p194. Cockle 531 & 932. Jähns.

FURTTENBACH, Joseph (1591-1667)

Halintro-Pyrobolia. Beschreibung Einer newen Büchsenmeisterey nemlichen Gründlicher Bericht wie der Salpeter Schwefel Kohlen vnnd das Pulfer zu prepariren zu probieren auch langwirrig gut zu behalten: Das Fewrwerck zur Kurtweil vnd Ernst laboriren. [*Description of a new system of gunnery and thorough instruction for the preparation of gunpowder from saltpetre, sulphur and charcoal, and for testing and preserving it, either for amusement or more serious purposes.*]

F130.1
HM1386/£650 Jonam Saurn: Ulm M.DC.XXVII. (1627)
1387/£250 EngTP+TP(BR)+Tab+xviii+108+44Pls. 31cms.
NF 0434801 BL 534.m.19 +

Architectura Vniversalis ...

F130.2
Johann Sebastian Medern: Ulm M.DC.XXXV (1635)
TP(BR)+xxiv+160+60Pls. 31cms. NF 0434797 BL 715.k.18

Buchsenmeisterey-Schul; Darinnen die newangehende Buchsenmeister ind Feuerwerk ... [*The school of artillery; a gunner's introduction to fireworks.*]

HM1388 £3600

F130.3
Johann Schultes: Augspurg M.DC.XXXXIII. (1643).
EngFPl+TP(BR)+EngPl+xii+151+44Pls. 31cms.
NF 0434799 BL 1265.f.25(2)

Mechanische Reissladen Das ist ein gar geschmeidige bey sich verborgen tragende Laden ...

F130.4
Johann Schultes: Augspurg 1644
NF — BL —

Mannhaffter Kunst-Spiegel ...

F130.5
1389/£420 Johann Schultes: Augspurg M.DC.LXIII. (1663)
TP(BR)+xxiv+292+32Pls. 31cms. NF 0434803 BL 1265.f.25(l)

Sotheran says: 'The Halinitro-Pyrobolia is the original form of what became afterwards, in a greatly enlarged form, the Buchsenmeisterey-Schul. The first part of Architectura Universalis treats of Fortification and illustrates and describes barracks and other military buildings ... the fourth part deals very fully with firearms and pieces of artillery, bombs and fireworks.' Although the text in Buchsenmeistery has been enlarged the same 44 plates are included as in the Halinitro-Pyrobolia. These show very clearly (see Illustrations) how a typical display of the period was setout - not so very different from today's layout. The Mannhaffter consist of 16 sections - 'Arithmetica, Geometrica ... Feurwerck, Buchsenmeistery' and the 4 'Architectura' (Militari, Civilii, Navali, & Insulata). The firework section occupies pages 180-208.

References: Fahler. Cockle 692 & 824. Du Pont. Berlin 3294 (1627). CP (1627)

GABARRET, Luciano

Manual de pirotecnia. (Selecta enciclopedia practica).

G010.1
Editorial Sintes: Barcelona (1953)
TP+196p+illus. 18cms. NG 0001573 BL —

Text and illustration plagiarised from IZZO.

References: Barfod

GALLERY

The Gallery and Nature of Art. Pyrotechny or the Art of Constructing Fireworks.

G020.1
London 1817
NG — BL —

References: Peddie

GAMES for all seasons. See **PARDON, G. F**

GANZE

Das Ganze der Feuerwerkerei; oder Anleitung zur Selbsverfertigung von Feuerwerken aller Art, sowohl zu Lande als zu Wasser; nebst Anhang für Artlleristen, von einem Kunstverständigen. [*All about making fireworks, or intructions for making yourself every kind of firework both for land and water, in addition to an appendix for gunners by one who knows the art.*]

G030.1
Wilhelm Lohmann: Goslar 1810
TP+iv+112 19cms. NG — BL —

References: Kayser. RDML

GEISSLER, Christoph Friedrichs von

Neue curieuse und vollkommene Artillerie ... Nebenst einem kleinen Anhang von Lust- Feuer- Wercken. [*New curious and perfect artillery ... to which is added a small appendix on recreative fireworks.*]

G040.1
Johann Christoph Zimmermann: Dresden 1718
TP(BR)+xiv+178+iv NG — BL 8834.i.7

Only the last ten pages are devoted to brief paragraphs describing rockets, saxons and water-fireworks.

References: RDML

GELINGSHEIM, Dr. Karl

Die moderne Kunstfeuerwerkerei. Eine Anleitung für Dilettanten. [*The modern firework maker. Instructions for amateurs.*]

G050.1
Strecker & Schröder: Stuttgart 1913
xi+222+8Tabs. 20cms. NG — BL —

References: GV. Barfod

GERLING, Walter

Fabricante pirotecnico.

G060.1

NG — BL —

References: R. Degn's 1971 catalogue

GILBERT, Charles

Endless Mirth and Amusement ... [*Including 'Fireworks and Chemical Surprises'.*]

G070.1
Dean & Son: London 1874
TP+ii+276. 19cms. NG 0208746 BL —

Fireworks and Chemical Surprises.

G070.2
Dean's Practical Guide Books: London 1879
TP+ii+46(-2). 19cms. NG — BL 8900.e.4

'Fireworks and Chemical Surprises' also bound into F.d'A Planches*Amusements without end ...*, Dean & Co: London 1874.

References: Stott 318 & 569

GILLET-DAMITTE, Jean Jacques Julien (1803-1875)

Bibliotheque usuelle des villes et des compagnes. I. Agriculture, Amendments et engrais. II. L'Art des feux d'Artifice, ou Pyrotechnie. III. Petit manuel de la bonne cuisine economique et simplifée. [*3 volumes in 1.*]

G080.1
C. Bleriot: Paris 1861

NG — BL —

Il Pirotecnico ossia l'Arte di Fare i Fuochi Artifiziali di Gillet-Damitte prima Versione italiana. [*Pyrotechnics or the art of making artificial fireworks.*]

G080.2
Seconda ed. C. Coen: Trieste 1865
TP+100(-2)+2FPls+Tab. 15cms. NG — BL —

G080.3
Terza ed. C. Coen: Venezia 1870
TP+TP+104(-2)+2FPls+iads. 16cms. NG — BL —

The BNC Catalogue lists dozen of text-books by GILLET-DAMITTE covering every conceivable subject, some of which ran to over 20 editions. The Italian version is presumably a translation of the 1861 edition.

References: Pagliaini gives edition of 1869. CP (1865 & 1870). BNC

GODE, Parasu-Rama Krishna

The History of Fireworks in India between A.D.1400 and 1900.

G090.1
Indian Institute of Culture: Bangalore 1953
26p. 25cms. NG 0266016 BL Ac.1929.C

India Office Library AC.1929.C. Transactions No.17 of Indian Institute of Culture.

References: CP

GOLDSMITH, R. B

Guide to Pyrotechnics. A Manual of Basic Methods.

G100.1
[Published by the Author]: Waupaca, Wisconsin 1953
34p (Typescript) 28cms. NG — BL —

One of a number of 'chap-books' issued in America after World War Two to encourage people to make their own in order to evade the restrictions placed on the sale of fireworks.

References: MS

GRAM, F

Lyst- Fyrvaerkeri for Dilettanter. Efter H. Leblanc. [*Fireworks for Amateurs. After H. Leblanc (q.v.)*]

G105.1
L.A. Jorgensen: Copenhagen 1866
NG — BL —

References: Vahl

GRIFFITHS, Capt. Frederick Augustus (? —1869)

The Artillerist's Manual, and compendium of ... Pyrotechny ...

G110.1
(1st ed.) J.M. Boddy: Woolwich (London) 1839
TP+x+300+Pls. 16cms. NG — BL 1140.b.21

The Artillerist's Manual and British Soldier's Companion of Infantry exercises ... Fireworks ... &c.&c.&c.

G110.2
2nd ed. For the Author. E. Jones: Woolwich 1840
TP+xxvi+372+12Pls. 15cms. NG 0521372 BL 1140.b.22

G110.3-10
3rd ed. 1843, 4th ed. 1847, 5th ed. 1852, 6th ed. 1864, 7th ed. 1856, 8th ed. 1859, 9th ed. 1862, 10th ed. 1868

In the first edition the section on 'Pyrotechny, Artificial Fireworks' occupies pages 224-238 and deals with squibs, leaders, Roman and blue candles, rockets, maroons, rain, jets, aerial globes, wheels, etc. The author's note reads, 'The following brief description of the requisite compositions and the methods of making Artificial Fireworks, will be considered perhaps extraneous in the present work, but the Compiler has been induced to insert it from his own personal knowledge of several instances in which Firemasters and Laboratory Serjeants have been called on to Superintend the construction, and exhibit a Display of Artificial Fireworks.'

References: CP (1840)

[GRIGNON & LALAING]

La Pyrothecnie practique, ou dialogue entre un amateur des Feux d'Artifice pour la Spectacle et un jeune homme curieux de s'en instruire ... ('Lettre du jeune comte de L ... a M. G ... amateur de la pyrotechnie.') [*Practical pyrotechnics, or a dialogue between a firework amateur and a young man eager to learn.*]

G120.1
L. Cellot & Jombert Fils: Paris 1780
340p. NP 0661846 BL —

GROTZ, Christopher

The Art of Making Fireworks, Detonating Balls &c. containing plain and easy directions for mixing and preparing the ingredients, and making and finishing the most simple devices in this ingenious art: together with how to make and fill air & fire balloons; and how to prepare and make detonating balls, spiders, segars, boots & shoes, Waterloo crackers, &c. The whole written expressly for this work, containing many devices never before published.

G130.1
Dean & Munday: London c.1819
ColFP+TP+22. 19cms. NG 0547492 BL 8715.aa.27

The complete art of making fireworks; improved to the modern practice; from the minutest to the highest branches; containing copious directions for making squibs, crackers, serpents, rockets, &c. ... by the Juvenile pyrotechnician.

G130.2
O. Hodgson: London 1819
ColFP+TP+22. 19cms. NG 0547495 BL 8715.aa.26

The art of making fireworks, improved to the modern practice ... with instructions for making squibs, crackers, serpents, wheels, Roman candles, rockets ... also for some highly curious aquatic fireworks.

G130.3
Thomas Richardson: Derby c.1820
ColFP+24p. 20cms. NG — BL —

G130.4
B. King: New York 1821
ColFP+28p. 19cms. NG 0547493 BL —

G130.5
B. King: New York 1822
ColFP+28p. 19cms. NG 0547494 BL —

G130.6
James Charles: Dublin (c.1821)
ColFP+TP+26. 19cms. NG — BL 8715.aa.46

G130.7
Willmer & Co.: Liverpool (c.1840)
24p. 12mo. NG — BL Mic.A.7191(6)

All of the above are virtually identical except for some variations in the long title. The Derby edition and the Hodgson edition do not contain the author's name but are assumed, at least by the NYPL, to be by Grotz due no doubt to their family resemblances. (See also 'The WHOLE secret laid open'). The engraved front cover of the Dublin edition has been reversed. Very primitive and vague descriptions. Another of the numerous 19th-century sixpenny chap-books.

References: Du Pont. NYPL. Rink 3183, 3185

GULDBERG See HØGH-GULDBERG

GUTTMANN, Oscar (1855-1910)

The Manufacture of Explosives. (2 Vols).

G140.1
Whittaker and Co.: London 1895

G140.2
(Specialist Series) Macmillan & Co.: New York 1895
(I)½T+TP+xii+348illus+7ads.
(II)½T+TP+x+444illus+10ads. 22cms.
NG 0416910 BL 08709.h.1

Die Industrie der Explosivstoffe.

G140.3
Fr. Vieweg und Sohn: Braunschweig 1895
ii+TP+xvi+704illus. 22cms. NG 0614908 BL 8905.ccc

Monumenta Pulveris Pyrii.

G140.4
For the Author. Artists Press: London MCMVI (1906)
½T+TP+vi+34+ii+102figs.+94Pls+iv. 34cms.
NG 0614913 BL LR271.b.4

Both the 1895 books contain the same substantial and important bibliography, (although lacking a lot of detail), which deals in the main with early books and manuscripts relating to gunpowder, explosives and related matters. The German edition forms part of the series of *Handbuch der Chemischen Technologie* edited by Dr. E. Engler.

References: Brock. CP (1895, 1895, 1906).

GUYOT, Gilles Edme (Membre de la Société Littéraire et Militaire Besançon)

Nouvelles Récréations Physiques et Mathématiques ... (3 Vols.).

G150.1
Chez Gueffier: Paris M.DCCLXXXVI (1786)
Troisieme Edition (I) ½T+TP+xiv+376+32Pls
(II) ½T+TP+iv+41O+48Pls (III) ½T+TP+408+iv+22Pls 20cms.
NG 0617179 BL 1608/2791

Eds. 1769/70. 1772/75. 1798/99. Volume II contains some 20 pages of a section called 'Recreations sur le Feu, l'Air et L'Eau' devoted not to 'real' fireworks but to the 'Maniere d'imiter au naturel les feux d'artifices réels, par la seule interposition de la lumiere et de l'ombre.' In general this consists of placing a light behind sheets of moving paper which have had designs painted on or pin-pricked. Such methods had many imitators, including HOOPER.

References: CP (1786)

H, R See **HOWLETT, Robert**

HALL, Trevor H

Mathematicall Recreations. An exercise in Seventeenth Century Bibliography.

H010.1
University of Leeds: 1969
½T+TP+ii+38. 21cms. NH — BL —

Limited edition of 200. Although not strictly a book about Fireworks, it nevertheless contains interesting and detailed analysis of the authorship of *Mathematicall Recreations* and *Récréation Mathematique*, as well as the bibliograhic inter-relationship of APPIER, ETTEN, LEURECHON, MALTHUS & OUGTHRED. Other titles which are of interest to bibliographers are his *Bibliography of Books on Conjuring from 1580-1850*, Lepton 1957 and *Old Conjuring Books*, London 1972. Both these contain a great deal of information about the large quantity of such books that also include material on fireworks.

References: CP

HANDBOK

Handbok i Vatten-, Teater-, och Salongs-Fyrverkeri. [*Handbook of water, theatre and indoor fireworks*]

H020.1
Sigfrid Flodins: Stockholm (1871)
PC+TP+ii+64+iv. 18cms NH — BL —

References: Barfod

HANZELET See APPIER, Jean

HARTMANN See RUGGIERI, C.F

HARZER, Friedrich See BUETTNER, F.C.A

HENRION, Denis (Pseud. of Cyriaque de Mangin). See MYDORGE, Claude

HERMART, Stefano

Il nuovo pirotecnico ossia l'arte di prepare i fuochi d'Artifizio. [*New pyrotechnics or the art of making firworks.*]

H030.1
Cioffi: Milano 1889
157p. 16mo. NH — BL —

References: Pagliaini

HERPIN, Jean Charles (1798-1872)

Recreations Chimiques, ou Recueil d'Experiences Curieuses et Instructives ... [*(2 vols.) Chemical recreations, or a collection of curious and instructive experiments.*]

H040.1
Audot: Paris 1824
384+304p+4FPls. NH 0313845 BL —

Recreaciones quimicas ... Traducidas del frances al espanol por D.A.G. y D.L.B. [*2 vols. in 1.*]

H040.2
F. Ifern: Barcelona 1827
NH 0313844 BL —

Amusements de Chimie, précédes de quelques appareils et de la maniere de les construire économiquement. Extraits en Partie des Recreations du Docteur Herpin.

H040.3
Rene: Paris 1842
230+FPl. NH — BL —

See also under Henri LAFONTAINE

HIME, Lieut. Col. Henry William Lovatt (1840- ?)

Gunpowder and Ammunition. Their origin and progress.

H050.1
Longmans, Green and Co.: London 1904
½T+TP+viii+256(-2). 22cms. NH 0378981 BL 8822.f.26

The Origin of Artillery.

H050.2
Longmans & Co.: London 1915
½T+Pic+TP+ii+232. 22cms. NH 0378984 BL 08821.h.25

There are only a few passing references to fireworks and two formulae are quoted, one from Reinaud and the other from Kentish. The author's note reads: 'Parts I and III of this book are a second and revised edition of *Gunpowder and Ammunition*, now out of print. Much of Part II has been taken from my article entitled *Our Earliest Cannon* which appeared in the Journal of the Royal Artillery Vol.XXXI.'

References: Brock

HOCUS POCUS

(Hocus pocus Junior). Hocus Pocus; or, the Whole Art of Ledgerdemain, Laid Open. Set forth in its proper Colours, fully, plainly and exactly; so that an ignorant Person may thereby learn the full Perfection of the same, after a little Practice ... To which is added, not in any former Edition, Instructions for Composing and Displaying several most beautiful pieces of Fireworks.

H060.1
James Bryn and T. Wilkinson ... : Dublin c.1760
143+i. 15cms. NH 0420256 BL —

H060.2
T. Wilkinson ... : Dublin c.1774
143+i. 15cms. NH 0420257 BL —

Trevor Hall suggests dates of c.1775 and c.1780. There is also an abbreviated entry in NUC under NH0420257 which gives a Dublin edition with a firm date of 1765. Is this another edition and does it contain the firework section?

References: Hall 169.3/.7. Stott 374/5

HODGETTS, Edward Arthur Brayley (1859-1910)

The Rise and Progress of the British Explosives Industry.

H070.1
Whittaker & Co.: London 1909
½T+Photo+TP+viii+418. 25cms. NH — BL 8903.h.2

Hodgetts was the editor of an interesting series of essays outlining the history of the British explosives industry up until the end of the 19th-century. There are articles by Hodgetts and G.W. MacDonald on Gunpowder, Col. J.R.J. Jocelyn on 'Military Fireworks', Phillip Pain on 'Pleasure Fireworks' as well as great detail concerning firms associated with the explosive industry at the turn of the century.

There is also a 10 page bibliography on Explosives as well as an historical 'Chronology'.

References: CP

HODGKIN, John Eliot

RARIORA: Being Notes of some of the Printed Books, Manuscripts, Historical Documents, Medals, Engraving, Pottery etc. etc. [*3 vols.*]

H080.1
Sampson Low, Marston & Co.: London [1902]
3 vols. 29cms. NH 0422766 BL 11908.h.16

A most remarkable catalogue compiled by of one of the last great 'amateur' collectors of the Victorian era. The range of subjects and items collected and described was vast; an impossible task today. What is particularly significant is his Firework Bibliography and collection of prints all fully, carefully and knowledgeably described. Many of his comments are quoted in this Bibliography. His whole collection was sold at Sotheby's in 1914, (his entire collection of some 70 Firework books being sold for just over £100!) many of the books going to the great bookseller Sotheran, whose Catalogues of the 1920s and 30s are a mine of useful information and critical comment.

References: CP

HØEGH-GULDBERG, C

Praktisk Haandbog i Lystfyrvaerkeri; til Brug saavel for Fyrvaekere som Dilettanter. [*Practical manual for Fireworks; for the use of Fireworkers as well as for Amateurs.*]

H090.1
(First part) I.H. Schubothes Boghandel. G.S. Wibes: Kjøbenhavn 1865
TP+vi+127 17cms. NH — BL —

H090.2
(Second part) Chr. Steen: Kjøbenhavn 1867
TP+vi+144+1FPl 17cms. NH — BL —

He started the regular series of firework displays in the Tivoli Gardens in Copenhagen in 1843 which continue to this day.

References: Vahl. Barfod

HOLIDAY

Holiday Frolics; or, Endless Amusements for Christmas Fireside. Containing ... Entertaining Experiments in various Branches of Science; Tricks with Cards and Dice, Art of Making Fireworks. The Whole admirably calculated to beguile the leisure hours of our Holiday Friends.

H100.1
W. Strange ... G. Purkess and J. Clements: London [1830]
ColFPl+36(-4). 15cms. NH 0458042 BL —

References: Stott 387. Hall 178.5

HOOPER, William

Rational Recreations.

H110.1
L. Davis: London MDCCLXXIV (1774). (4 vols.)
Vol.I ½T+TP+xvi+268+12FPls
Vol.II ½T+TP+xii+280+19FPls
Vol.III ½T+TP+xii+296+20FPls
Vol.IV ½T+TP+xii+368+14FPls. 21cms. NH 0508075 BL 49.b.20

H110.2-5
2nd ed. cor. 1782/83, 3rd ed. cor. 1787, 4th ed. cor. 1794 1801(2 Vols.) and 1802(2 Vols.)

Much of this work stems from GUYOT. In the section devoted to *Pyrotechnics* he says (Vol.IV. p156), 'We shall here add the description of a new method of imitating artificial fireworks, which appears to be the invention of the ingenious M. Guyot.'

References: Hall 180-4. Stott 389-394. CP (1774)

HOSOYA, Masao (1911-)

Hanabi no kagaku. [*The Science of Fireworks*]

H120.1
Tokai University Press: Tokyo 1980
173p. 18cms. NH — BL —

The author is the owner of the Hosoya firework factory and has had some 60 years' manufacturing experience. The book, which is written for students and scholars, consists of five chapters which are devoted to: The history of fireworks, The development of fireworks, Types of firework composition, The science of fireworks and miscellaneous items. It also contains a bibliography of Chinese firework books.

References: CP

HÖTTINGER, Ferdinand See BUETTNER, F.A.C

HOWLETT, Robert

The School of Recreation: or a guide to the most ingenious exercises ... of Fireworks ...

H130.1
(1st ed.) H. Rodes: London 1684

H130.2
H. Rodes: London 1696
Eng+TP+iv+182. 15cms. NH 0563892 BL 1651/1126

H130.3-8
Similar eds. H. Rodes: 1701. A. Bettersworth: 1710, 1719, 1720, 1732, 1736

The first edition of 1684 did not contain a section on fireworks. The first to have a firework section was that of 1696, where it occupies pages 26-41. The subjects dealt with include, sky rockets, golden rain, silver stars, red stars, balloons (mortars and shells), saucissons, girondels, St. George and the Dragon, wheel of fireworks etc. The other sections in this collection include: Hunting, Riding, Racing, Military Discipline, The Science of Defence, Hawking, Tennis, Bowling, Ringing, Singing, Cock-Fighting, Fowling & Angling. The author in his introduction has this to say of his curious selection, 'I commend them as they are very suitable recreations for the Gentry of England ... '

References: Wing H3194/5. Arber (1684, 1696).

HUTSTEIN, J See WEBSKY, Martin

HUTTON, Charles (1737-1823)

Recreations in Mathematics and Natural Philosophy. First composed by M. Ozanam ... lately recomposed and greatly enlarged in a new edition, by the celebrated M. Montucla. And

now translated into English, and improved with many additions and observations, by Charles Hutton. [*4 vols.*]

H140.1
T. Davison for G. Kearsley: London 1803
Vol.I ½T+TP+xvi+447+21FPls
Vol.II ½T+TP+xiv+464+34FPls
Vol.III ½T+TP+xii+502+33FPls
Vol.IV ½T+TP+xii+516+8FPls. NO 0189044 BL 537.h.13/16

H140.2
Longman, Hurst etc.: London 1814
NO 0189045 BL 1500/70

Recreations in Mathematics and Natural Philosophy: tr. from Montucla's edition of Ozanam by Charles Hutton.

H140.3
A new & rev. ed. by Ed. Riddle. T. Tegg: London M.DCCC.XL (1840)
xiv+826illus. 23cms. NO 0189046 BL 1135.k.17

H140.4
The present new edition by Ed. Riddle. T. Tegg: London 1844
xiv+826illus. 23cms. NO 0189047 BL —

H140.5
New ed. rev. & cor. by Ed. Riddle. W. Tegg: London 1851
xiv+826illus. 23cms. NO 0189048 BL —

H140.6
New ed. rev. & cor. by Ed. Riddle. W. Tegg: London 1854
xiv+826illus. 23cms. NO 0189049 BL —

The section on Pyrotechny occupies pages 437-502 in Vol.III of the first edition of 1803 and includes two plates. Charles Hutton translated this work from the revised version made by Montucla of OZANAM's *Récréations Mathématiques et Physiques*. (1778 ed.). As Hutton says in the opening to the Pyrotechny section, 'Why it has been usual to consider pyrotechny as a branch of the mathematics, we do not know ... However, as we might be blamed for omitting an art which affords a considerable field for amusement, and as it is connected at least with natural philosophy, we shall make it the subject of one division of this work.' It deals with gunpowder, various different sizes of rockets, Chinese fire, serpents, maroons, saucissons, line rockets, aerial globes called bombs, jets and coloured fire, as well as a few pages on 'imitation artificial fireworks' which he calls 'optical pyrotechny'. He commends the works of FREZIER and PERRINET d'Orval to those who are 'fond of the Art'. Sotheran describes the 1803 edition as the 'best English edition of this popular work'.

References: SOT. Hall 215. Stott 521-524

INCARVILLE, Pierre d' (1706-1757)

Maniére de faire les fleurs dans les Feux d'Artifice Chinois. [*The method of making flowers in Chinese fireworks.*]

I010.1
Mémoires de l'Academie des Sciences: Paris 1763
Tome IV pgs.66-95. NA0 036486 BL 985.b.1-13

The Encyclopaedia Britannica (4th ed. 1810) refers to a 'Manuel d'Artificier published at Paris in 1757 by Father d'Incarville' . Backer says, 'Manual de l'artificier, ou se trouvent les compositions chinoises du P. d'Incarville, pour representes les fleurs de diversée éspèces. Paris Jombert 1758 (sic) 12mo'. They probably refer to the version published by Jombert which contains Incarville's work plus that of FREZIER and PERRINET d'Orval. The above article was subsequently translated into English and printed in the *Universal Magazine* for January 1764 (Page 21) under the title 'Manner of making flowers in the Chinese Fireworks, illustrated with an engraved copper-plate - From the fourth volume (just published) of the Memoires presented to the Academy of Sciences.'

The NUC listing is under 'ACADEMIE des Sciences: Paris', the BL listing is under 'PARIS: Academie des Sciences.'

References: Backer

ISACCHI, Giovanni Battista da Reggio

Inventioni ... nelle quali si manifestano varij secreti, et utili avisi a persone di guerra, e per i tempi di piacere ... [*Inventions ... in which are revealed various secrets and useful information for military use and in times of peace.*]

I020.1
Seth Viotto: Parma 1579
xvi+172+vii. 21cms. NI 0166513 BL 58.a.4

Contains some 60 *Secrets and Inventions* by various named persons including about 10 dealing with 'Girandole', 'Trombe de Fuoco artificiale', 'Fulmine di Fuoco' etc. Numerous quaint woodcuts. Hodgkin says, 'This is in some respects one of the most remarkable books on Fireworks that I possess. I have never seen another copy, nor is the book described in any work I have consulted. Its author was evidently a man of most independent thought and his mechanical schemes are startlingly original ... He is keenest, as all the early writers were, on Fireworks for Military purposes, fire-ships, fire-lances, fire-swords, fire-axes, but has also something to say about recreative pyrotechny ...'

References: Cockle 547

IZZO, Colonello Dott Attilio (1898- ?)

Pirotecnia e Fuochi Artificiale, manuale pratico e ricettario per chi fabbrica fuochi artificiali e chi vende materie prime per usi pirotecnici. [*Pyrotechnics and artificial fireworks, practical manual and formulae for those who make artificial fireworks and those who sell raw materials for pyrotechnic use.*]

I030.1
Hoepli: Milano 1950
xvi+274illus. 19cms. NI 0199176 BL —

Although in the preface the author indicates that this book is a revision of Fiorina di MAIO's *Pirotechnia moderna'* about two-thirds of it is actually taken from it verbatim. He wrote a number of other works concerned with chemical warfare and explosives. See also GABARRET.

References: NYPL

JOCELYN, Julian Robert John (1852- ?)

The connection of the Ordnance Department with National and Royal fire-works, including some account of Sir Martin Beckman, Colonel Henry John Hopkey, & Sir William Congreve.

J010.1
Royal Artillery Institution: Woolwich Feb. 1906
29p. illus. 25cms. NJ — BL AC.4357

This interesting article, showing how closely connected with recreational fireworks were the gunners and officers of the Royal Ordnance, appears from page 481 of the February 1906 edition of the *Royal Artillery Institute Journal* (Volume XXXII of the bound volumes). From the early part of the 17th century, 'the provision of national or Royal fireworks was the duty of the Master General of the Ordnance and specialists in pyrotechny became a necessity of the Department. Firemasters as distinct from Master Gunners began to fill regular appointments; in

1672 a regular Laboratory "for receiving fireworks" was commenced at Woolwich'. This work continued until 1856. A small forensic laboratory is still there today.

References: Du Pont. Brock

JONES, Lieut. Robert R.A

A New Treatise on Artificial Fireworks.

J020.1
Printed for the Author; and sold by J. Millan: London MDCCLXV (1765)
TP+xxvi+255+5FPLs. 20cms. NJ 0155527 BL 07945.f.24

Artificial Fire-Works, Improved to the Modern Practice. From the Minutest to the Highest Branches; ... The Second Edition, with the Addition of many new and beautiful Fire-Works, and 3 large Copper-Plates.

J020.2
J. Millan: London MDCCLXVI (1766)
TP+xxvi+262+8FPls. 21cms. NJ — BL Cup.401.c 37

JONES, Capt. Robert

Artificial Fireworks, Improved to the Modern Practice, From the Minutest to the Highest Branches; ... The Second Edition, corrected. With the addition of many new and beautiful Fire-Works, and three large Copper Plates. ... Also, Mr. Muller's Fireworks for Sea and Land Service ...

J020.3
J. Millan: London MDCCLXXVI (1776)
TP+iv+194+8FPls+3FPls. 21cms. NJ 0155525 BL 1420.c.4

Artificial Fireworks, Improved to the Modern Practice, From the Minutest to the Highest Branches; ... with all their Ingredients, Compositions, Preparations, Machines, Moulds, and manner to make them; ...

J020.4
Meggy and Chalk: Chelmsford 1801
TP+iv+184+20Pls(10ds). 19cms. NJ 0155526 BL 8715.aaa.45

The first edition is remarkable as it is the first book dealing exclusively with recreational fireworks in English since BABINGTON's classic work of 1635, one hundred and thirty years earlier. The second edition is a reissue of the original edition of 1765 with a cancel titlepage, and pages 253-255 excised from the original with the additional 11 pages from 253 to 262 describing the set-pieces depicted on the extra 3 Plates. Even the 'Errata' noted on page xxvi of the first edition have not been corrected. The contents of the 1776 edition are similar to that of the 1766 having only been corrected and reset. Between the second edition of 1766 and the 'corrected', and indeed edited and abridged, second edition of 1776 Jones was evidently promoted to Captain. The three additional plates in this 1776 edition are associated with Belidor's chapter on Mining which is not mentioned on the Title page. The ECB lists an 1802 edition entitled 'The whole art of Making Artificial Fireworks' published by Hurst. I have not seen any further details. He also wrote a number of books on skating!

References: Du Pont (1766). Brock. CP (1765, 1766, 1776, 1801)

KENTISH, Thomas

The Pyrotechnist's Treasury; or, Complete Art of Making Fireworks.

K010.1
Chatto & Windus: Piccadilly, London 1878
½T+TP+viii+2OO+16Pls+xxxiiads. 19cms. NK 0098712 BL 8716.aa.49

The Pyrotechnist's Treasury; the Complete Art of Making Fireworks. 2nd ed. rev. and enl.

K010.2
Chatto & Windus: Piccadilly, London 1887
½T+TP+viii+200+16Pls. 19cms. NK 0098714 BL —

The Complete Art of Firework Making; the Pyrotechnist's Treasury. A new edition.

K010.3
Chatto & Windus: Piccadilly, London 1899
xiv+218+24Pls. 19cms. NK 0098711 BL —

The Pyrotechnist's Treasury. The Complete Art of Firework-Making.

K010.4
Chatto & Windus: London 1905
½T+Pl+TP+x+218+24Pls. 19cms. NK 0098715 BL —

K010.5
(Reprint) Paladin Press: Boulder, Colorado. c.1976
xiv+218+plates. 22cms. NK — BL —

Despite having first been printed over 100 years ago this remains one of the greatest classic how-to-do pyrotechnic books that has ever been written. Much of its detailed description and practical advice is as relevant today as it was in the latter part of the 19th century. The reviewer of the reprint, *American Pyrotechnist Fireworks News* September 1976, said, 'One could well believe that he is reading a recently-published new book on fireworks, except that none printed in the intervening 71 years are so concise, detailed and painstakingly illustrated'.

References: Du Pont. Brock. NUC SC 1977. CP (1878, 1905)

KLERCKER, G. af Major

Handbok i Lust-Fyrverkeriet för Artillerister och Konstälskare. [*Handbook of recreative fireworks for artillerymen and lovers of the art.*]

K020.1
F.F. Cedergreens: Christianstad 1829
PC+TP+ii+201+3Pls. 19cms NK — BL —

References: Barfod

KOLOMNII, V

Obshchedostupnaya pirotekhiya. [*Pyrotechny for everyone.*]

K030.1
1880
NK — BL —

References: SHIDLOVSKY's bibliography

KORT

Kort Afhandling om Lustfyrwerkeriet, für tillwerkandet af de enklaste och mest wanliga pjefer. [*A short treatise of recreational fireworks, for the manufacture of the simplest and most common pieces.*]

K040.1
Stockholm 1847
NK — BL —

K040.2
Isaac Marcus: Stockholm 1857
PC+22p. 16cms. NK — BL —

References: CP (1857).

KORTFATTET

Kortfattet Anviisuing til at forfaerdige Lystfyrvaerkerier. [*Brief instructions for making fireworks.*]

K050.1
H.P. Møller: Copenhagen (1854)
NK — BL —

References: Barfod

KUNST

Kunst der Feuerwerkerei. Anleitung zur Anfertigung sammtlicher Feuerwerksobjecte, als aller Arten von Luchtfeuern, Sternen, Leuchtkugeln, Raketen, der Luft- und Wasserfeuerwerke. [*The art of making fireworks. Instructions for the making of all firework items as well as all kinds of lights, stars, fireballs, rockets, air and water fireworks.*]

K060.1
U. Hartleben: Wien, Pest, Leipzig [1885]
½T+TP+171(45illus.). 17.5cms. NK — BL 8716.aaa.13

References: Du Pont. Kayser. G&O (1885).

L-e, L. v

Der Wiener Kunst- und Lustfeuerwerker … dargestellet durch L. v. L-e … [*The art of Viennese fireworks.*]

L010.1
Carl Gerold: Wien 1818
TP+xxx+348+30Pls. 20cms. NL — BL —

Vollständiges Taschenbuch für Kunst und Lustfeuerwerker und Liebhaber dieser Unterhaltung. [*Complete pocket-book for artificial and recreative firework making and lovers of the entertainment.*]

L010.2
Hartleben: Pest 1820
TP+xii+346+27Col.Pls. 20cms. NL — BL —

The coloured plates are outstanding and detailed.

References: Lotz. Berlin 3308. RDML (1818, 1820)

LAFONTAINE, Henri

Manuel Illustré des Amusements de Chemie. Précédé de quelques appareils la maniére de les construire economiquement. [*Illustrated manual of chemical amusements.*]

L020.1
Passard: Paris c.1830
½T+TP+230(-4)+1FPl. 14cms. NL — BL —

The half-title adds 'Bibliothéque des Connaissances Utiles'. The last 44 pages are headed 'Pyrotechnie ou Feux D'Artifice'. The wording and the collation are virtually identical to the book listed under HERPIN. However a direct comparison between the two has not been possible.

References: CP

LALANNE, (Marie) Ludovic (Chrétien) (1815-1898)

Essai sur le feu gregois et sur l'introduction de la poudre a canon en Europe, et principalement en France. [*Essay on Greek fire and the introduction of gunpowder into Europe.*]

L030.1
Bachelier: Paris 1842. (Extrait des Annales de Chimie et de Physique. 3rd series T.IV)
21p. NL — BL —

L030.2
Academie des Inscriptions et Belles-lettres: Paris 1843
2nd series Vol.1. p294-362. NL 0040750 BL AC420/10(2)

Recherches sur le feu gregois et sur l'introduction de la poudre a canon en Europe. 2nd ed. cor. et entièrement refondue.

L030.3
J. Correard: Paris 1845
96p. NL 0040762 BL 1398.k.8

References: Brock. BN

LAMARRE, A

Nouveau Manuel de l'Artificier, ou Traité practique pour le fabrication des feux de couleurs, des torches Bengale, et de tous les artifices de joie. [*New manual of fireworks, or practical treatise for the manufacture of coloured fires, Bengal torches and all kinds of fireworks.*]

L040.1
J. Baudry. Chez Auteur: Paris 1878
½T+TP+188(incl.29figs.). 18cms. NL 0044784 BL 8716.b.7

Michael Swisher comments that 'this is a curious work as the author advocates the use of linseed oil as a fuel in pyrotechnic compositions. There is a catalogue of pyrotechnic chemicals for sale by the author in the back'.

References: Brock

LAMMAR, A

Feyerverochnoye iskusstvo. [*The fireworker's art.*]

L050.1
St. Petersburg 1914
NL — BL —

References: SHIDLOVSKY's bibliography

LANCASTER, Rev. Ronald M.A. (1931-)

Fireworks. Principles and Practice.

L060.1
Chemical Publishing Co. Inc.: New York 1972
TP+viii+274incl.134figs. 22cms. NL — BL —

Originally conceived by the author to be a 'revision' of WEINGART'S classic work *Pyrotechnics* it became, in practice, a completely fresh work incorporating all the developments in firework making that have occurred in the 25 years since the appearance of WEINGART's second revised edition of 1947. The three other co-authors are: Roy Butler, who contributed a short historical section, R.G. Hall, on the use of polymerising plastic resins, and Takeo SHIMIZU whose translation from his *Hanabi* on Japanese shell making is the first detailed English treatise of the subject. Together with WEINGART and KENTISH it is assured a place as one of the most practical works on the subject in the 20th century. It was reviewed in the January 1972 edition of *American Pyrotechnist Fireworks News.*

References: CP

LEBLANC, Henri

Der vollkommene Lust-Feuerwerker, oder: Gründliche und leichtfassliche Anleitung, in kurzer Zeit ohne grossen Kosten-Aufwand die Schönsten und überraschendsten Feuerwerkstüke, als: Sonnen, Glorien, ... wie überhaupt vollständige Zimmer-, Land-, Theatre- und Wasserfeuerwerke auf eine leichte Art anzufertigen. Zur Berschönerung öffentleicher und häuslicher Feste. Nach eigene pracktischen Erfahrungen und den exprobten Bortheiten der berühmtesten Feuerwerker für Dilettanten bearbeitet von Henri Leblanc. [*The complete fireworker, or; basic and simple instruction for the making of the most beautiful and surprising fireworks easily and cheaply.*]

L070.1
(1st ed.) B.S. Berendsohn: Hamburg 1863
TP+x+172+8Pls. 13cms. NL — BL —

L070.2
Zweite Aufl. B.S. Berendsohn: Hamburg 1864
TP+x+172+viiiads.+8Pls. 13cms. NL — BL 8715.a.9

L070.3
(3rd ed.) B.S. Berendsohn: Hamburg 1867
TP+x+172+8Pls. 13cms. NL — BL —

WEINGART's bibliography lists the 1863 edition under Berendsohn, who is the publisher of this work. See also GRAM.

References: Kayser

LEHALLEUR, Jean Pepin

Traité des poudres explosives et artifices. [*Treatise of explosives and fireworks.*]

L080.1
Bailliere: Paris 1935
NL — BL —

LEURECHON, Jean See ETTEN, Henri van

LINDER, Carl

Praktisk Handbok i Lustfyrverkerikonsten. [*Practical handbook for fireworkmakers.*]

L090.1
Sigfrid Flodins: Stockholm (1871)
PC+TP+viii+144inc.41figs. 18cms. NL — BL —

L090.2
Stockholm 1880
NL — BL —

References: Barfod

LODEN, Arthur

Der Lustfeuerwerker. Eine für Dilettanten fassliche Anleitung zur Lustfeuerwerkerei. [*The fireworker. Instruction for amateurs to make fireworks.*]

L100.1
Ernst'schen Verlag: Quedlinburg 1840
130p. NL — BL —

Der Lustfeuerwerker oder gründliche Anweisung zur Lustfeuerwerkerei.

L100.2
(New TP) Ernst'schen Verlag: Quedlinburg und Leipzig 1849
130p. NL — BL —

L100.3
(2nd ed.) Ernst'schen Verlag: Quedlinburg und Leipzig 1857
130p. NL 0438691 BL —

L100.4
(3rd ed.) Ditto? 1860
NL — BL —

L100.5
Vierte Aufl. Ernst'schen Verlag: Quedlinburg und Leipzig 1862
TP+iv+138+1FPl. 15cms. NL — BL —

L100.6
(5th ed.) Ditto? 1870
NL — BL —

L100.7
Sechste Aufl. Ernst'schen Verlag: Quedlinburg und Leipzig 1872
TP+vi+158+2FPls. 17cms. NL — BL —

L100.8
(7th ed.) Ditto? 1875
NL — BL —

L100.9
(8th ed.) Ditto? 1880
NL — BL —

L100.10
(9th ed.) Ernst'schen Verlag: Quedlinburg und Leipzig 1885
viii+157+1FPl. 17cms. NL — BL —

L100.11
(10th ed.) Ditto? 1898
NL — BL —

References: Du Pont (9th ed.). Keyser. G&O (1885). CP (1862, 1872)

LOTZ, Arthur

Das Feuerwerk, seine Geschichte und Bibliographie. [*Fireworks, a history and bibliography.*]

L110.1
Karl W.Hiersemann: Leipzig (1941)
TP+iv+136+36Pls. 32cms. NL 0503657 BL —

L110.2
(Reprint) Edition Olms: Zurich 1978
½T+TP+ii+TP+iv+136+36Pls. 28cms. NL — BL X0622/534

Reviewed by Michael Swisher in *Pyrotechnica* No.VII. Whereas BROCK's works deal with the overall history of fireworks, both technical and historical, the LOTZ work concentrates on the artistic and architectural elements, specifically from the sixteenth to the eighteenth century, and the connection between the great European and Russian ceremonials and celebrations of the period, and the artists concerned in the design of firework displays that were associated with them.

References: CP (1978).

LUCAR, Cyprian

Three Bookes of Colloqvies concerning the Arte of Shooting in great and small pieces of Artillerie ... written in Italian, and dedicated by Nicholas Tartaglia ... And now translated into English by CYPRIAN LVCAR ... Also the said CYPRIAN LVCAR hath annexed vnto the same three bookes of Colloqvies a Treatise named LUCAR APPENDIX collected by him out of diuers Authors in diuers languages to show vnto the Reader the properties, office, and dutie of a Gunner, and to teach him to make and refine artificial saltpeeter, to sublime brimstone for gunpowder to make coles for gunpowder, to make gunpowder of diuers soretes and of diuers colours, to make gunmatches, ... to make and vse diuers Trunkes, and many sortes of fireworkes ... and to do other commendable things which not onelei in time of warre, but also in time of peace may to a good end be practised.

L120.1
By Thomas Dawson for Iohn Harrison: London 1588
TPEng+vi+80+TPEng+ii+120+ii+4Pls. 28cms. NT 0044435 BL 1651/761

The 'Appendix' which has a separate title-page, consists of 113 Chapters devoted to all the subjects indicated by the long-title. However the distinction between the military use and the peaceful use of fireworks is very blurred. Only three or four of the chapters specifically refer to 'Triumphes'. For example, Chapter 90: 'To make a fireworke which may be used in militarie service, and also in triumphes'. Chapter 92: 'To make a fireworke which at a triumph may be cast up into the ayre'. The same composition having been used in a previous chapter, fired from 'Trunckes' (i.e. tubes like a Roman Candle) at the enemy to cause 'mischieve'. In his introduction to the 'Appendix' the author acknowledges the numerous authors whose works he has 'collected from'. These include 'Vannucio Biringuccio,

Girolamo Ruscelli, Girolamo Cataneo, Francesco Ferretti, Cosimo Bartoli, ... Luigi Collado, ... Daniel Santbach, ... Ioan: Baptista Porta, Hieronumous Cardanus, ... Thomas Finck, ... M. Robert Recorde, M. William Cunningham, M. Leonarde (Thomas) Digges, M. Peter Withorne [Whitehorne].

References: Cockle 38 & 39

M, D See MANLYN, Daniel

MAGAZINES

Zeitschrift für Pyrotechniker aller Art, als Feuerwerker, Pulvermüller, Ingenieure, Salpeter und Schwefelarbeiter, Verfertiger neuester Feuerzeuge und der Zündhütchen. [*Journal for every kind of pyrotechnist; such as fireworkers, powder millers, engineers, saltpeter workers, and makers of firelighters and matches.*]

M010.1
B.J. Voigt: Weimer 1844-55

The above publication was issued in three volumes, each consisting of six parts, and each part containing 32 pages. The parts appeared approximately every nine months from January 1844 until 1855. The first volume was edited by Dr. Christian H. Schmidt and Alexander Schultz and volumes two and three by Dr. Carl Hofmann. Full details are given in Kayser. Reviews of books by UCHATIUS, WEBSKY, BUETTNER and SCHAFENBURG are included as well as details of the then new discoveries of CHERTIER and others.

Pyronews.

M010.2
(Editor) Max Vander Horck, California: November 1966 - December 1967. (13 issues. No December 1966)

American Pyrotechnist Fireworks Monthly.

M010.3
(Editor) Max Vander Horck, San Diego, California: January 1968 - September 1970. (24 issues)

Fireworks News.

M010.4
(Editor) P.N. Colonnese, Pyro Press Publications, Lexington: January 1970 - September 1970

American Pyrotechnist Fireworks News.

M010.5
(Editor) Max Vander Horck, Pyro Press Pubs., Lexington: October 1970 - November 1976

American Pyrotechnist.

M010.6
(Editor) Max Vander Horck: San Diego. January 1977 - June 1981
(No issue December 1978, 1979, 1980)

American Firework News.

M010.7
(Editor) J. Drewes: Rex E. & S.P. Inc., New Jersey & Whitewater, Colorado: July 1981 -

Bulletin of The Pyrotechnics Guild International Inc.

M010.8
Schaumburg, Illinois: 1974 -

Pyrotechnica - Occasional Papers in Pyrotechnics.

M010.9
(Editor) Robert G. Cardwell: Austin, Texas
I - October 1977. II - February 1978. III - May 1978
IV - October 1978. V - October 1979. VI - July 1980
VII - May 1981. VIII - June 1982. IX - June 1984

Fireworks.

M010.10
(Editor) John Bennett: Bexhill-on-Sea, England
Occasional from 1982

Green Notes.

M010.11
J. Finckbone: Megatech Supplies. U.S.A
Occasional from 1983

Pyronews was the first magazine to be published on the subject in English, by Max Vander Horck in November 1966. In January 1968 it became the *American Pyrotechnist Firework Monthly* which merged with *Fireworks News* in October 1970. It subsequently became *American Pyrotechnist* in January 1977 and then, on the retirement of Vander Horck in June 1981, it became *American Firework News.*

Whereas these magazines were essentially parochial in style, *Pyrotechnica* takes an international view and its articles are more technical and theoretical in approach.

References: CP. (1844-55 and all after 1970)

MAGIC

The Magic Oracle; or, Conjuror's Guide; containing the Whole Art of Legerdemain, the best and newest tricks with cards, dice etc. and directions for making Fireworks.

M020.1
Printed for the Booksellers: Glasgow c.1850
ii+24(-3). 16cms. NM 0112001 BL —

References: Stott 471

MAIO, Fiorina de

Pirotecnia Moderna ...

M030.1
U. Hoepli: Milano 1891
viii+150. NM 0133082 BL —

M030.2
Seconda edizione riv. et amp. Hoepli: Milano 1903
½T+TP(BR)+x+183(inc121fig). 15cms. NM 0133083 BL 012200.hh.65

M030.3
Tertza ed. riv. e amp. Hoepli: Milano 1916
xv+183. NM — BL —

M030.4
Quarta ed. riv. e amp. Hoepli: Milano 1930
xxi+268. NM — BL —

Manual de pirotecnia, par F. de Maio y M. Jona: version del Italiano por J. de D.S.H. 2nd ed. aum. con arregio a la 4e Italiana.

M030.5
G. Gili: Barcelona 1931
ii+288illus. 18cms. NM 0133084 BL —

The 1891 edition contains the first published detailed description of the 'Italian' style strung and pasted cylinder shell. See note under IZZO.

References: CUBI. Pagliaini. NYPL (1930). MS

MALTHE, Francois de

Traité des Fevx Artificiels povr la Gverre, et povr la Recreation; avec plusieurs belles obseruations, abregez de Geometrie, Fortifications & exemples d'Arithmetique ... par le sieur F.D.M. [*Treatise of fireworks for war and recreation.*]

M040.1
Pierre Gvillemot: Paris M.DC.XXIX (1629)
EngTP+TP+x+243+xi(illus). 18cms. NM 0156654 BL —

A Treatise of Artificial Fire-vvorks both for VVarres and Recreation: with divers pleasant Geometricall obseruations, Fortifications and Arithmeticall Examples. In fauour of Mathematicall Students. Newly written in FRENCH, and Englished by the Authour THO: MALTHVS.

M040.2
Richard Havvkins: Chancerie-lane (London) 1629
EngTP+TP+x+264+viii+30WC. 17cms. NM — BL 787.a.75

Traité des Fevx Artificiels ... De nouueau reueu, corrigé, & augmenté par l'Auteur ...

M040.3
Pierre Gvillemot: Paris 1632
EngTP+TP+vi+277. 18cms. NM 0156655 BL —

M040.4
Pierre Gvillemot: Paris M.DC.XXXIII (1633)
TP+vi+245+ix(inc.WC). 17cms. NM — BL 534.d.17

M040.5
Cardin Besogne: Paris M.DC.XXXX (1640)
EngTP+TP+viii+282+WC. 17cms. NM 0156656 BL —

M040.6
Cardin Besogne: Paris M.DC.LXI (1641)
vi+277+v(illus). 17cms. NM 0156657 BL —

Practique de la Guerre ... & a la fin un traite des Feux de Ioye.

M040.7
I. Gvillemont: Paris 1646
TP+292+2Pls. 22cms. NM 0156650 BL —

Practiqve de la Gverre ... Ensemble vn Traite des Feux de Ioye.

M040.8
Gervais Clovsier: Paris 1650
TP+EngTP+vi+282inc.26Pls+vi. 24cms. NM 0156652 BL 534.i.26(1)

Practiqve de la Gverre ... & à la fin un Traitté des Feux de Ioye.

M040.9
Gervais Clovsier: Paris M.DC.LXVIII (1668)
TP+x+252inc29Pls. 16cms. NM 0156653 BL 54.a.24

M040.10
Gervais Clovsier: Paris 1674
TP+x+252incl29Pls. 16cms. NM — BL —

M040.11
Gervais Clousier: Paris M.DC.LXXXI. (1681)
TP+x+292+29Pls. 18cms. NM — BL —

HODGKIN says of the English edition in his *Rariora*, 'A charming little volume ... His instructions for Firework making are far more voluminous. Rocketes, Serpents, Golden Rayne, Starres giving great reports, Saucissons, Fierie Boxes, Fierie Lances, Girondelles or Fierie Wheeles, Ballouns, and Flying Saucissons, all have their place in this category, and from the list we may obtain a very accurate idea of the development of the art at that period.' In this edition of 1629 page numbers 171/172 are used twice giving 264 pages not 262 as paged. Pages 1-60 are devoted to 'War', pages 61-135 to 'Recreation' and the remainder to 'Geometrie'. This English version appears identical to the French edition, with the same chapter and paragraph headings and virtually the same plates although they have been re-engraved.

Sotheran says of *Practique de la Guerre*, 'Especially interesting for its chapters on bombs, hand-grenades and Fireworks'. In all editions that I have seen the pages are misnumbered, pages 134-143 being repeated twice thus giving a total of 292 and not 282 as the last page is numbered.

References: Du Pont. Cockle 118 & 939. BNC (1632, 1646, 1650, 1668, 1672). Berlin 3296 (1632). STC17217. CP (1629, 1640, 1681). NYPL (1632, 1650)

MALTHUS, Thomas See MALTHE, Francois

MANGIN, Arthur

Délassements Instructifs. Les Télégraphes. Les Feux de Guerre. [*Instructive recreations.*]

M050.1
A. Mame: Tours 1855
188p. NM 0168583 BL —

M050.2
Deuxième ed. Ad. Mame: Tours 1856
½T+Eng+TP+188. 19cms. NM — BL —

The half-title reads 'Biblioteque des Ecoles Chrétiennes approuvés par S. Em. le Cardinal Archêveque de Tours'. The last 100 pages are devoted to 'Le Feux Gregois', 'La Poudre à canon' and 'Tentatives de reforme la pyrotechnie moderne'. Hardly suitable subjects for Christian schools one would have thought. There are seven editions and numerous re-impressions listed in the Bibliotheque Nationale Catalogue.

References: BNC. CP (1856)

[MANLYN, Daniel]

Pyrotechnia, of Konstige Vuurwerken, waer in wort seer duydelijek beschreven de manier van het maken en toestellen van allerley vuur-wercken tot vermaeck, hetzy loopende of staende, als mede een Tweede Deel, begrijpende een Onderrichtinge van het maken Vuurwercken tot Dienst, ... uyt verscheyde Autheuren getrocken en Vertaelt, eerste uitgave. (Pt.II) Pur erga ys latryan ofte Vuurwerken tot dienst. [*Pyrotechnia, or artificial fireworks, in which is very clearly described the manner of making and erecting all sorts of fireworks for amusement, both flying and fixed, as also a second part comprising a course in the making of fireworks for service ... excerpted and translated from various authors, first edition. 2 parts in 1*]

M060.1
(Pt.I) Hugo Rijckhals: Rotterdam 1672
EngTP+TP+iv+192+22Pls HM 2133 £420

M060.2
(Pt.II) Hugo Rijckhals: Rotterdam 1671
TP+152(-2)+viii+6Pls. NM 0174935 BL C.105.a.9

Pyrotechnia, of Konstige Vuur-werken, vertoont in vuurspouwende loopende Draken en Paarden, aardige Kasteelen, vechtende Schepen, geestige Lantarens, wonderlijke Vuurpijlen ..., Aerdklooten in de lucht, stralende Starren ... Nessens Een heerlijk van't maken van Vuurballen, Granaden, Storm-kranssen, Stick-potten, Petarden. Dienstigh voor alle jonge Constapels en andere Liefhebbers, die haer in de Oorlogs-konst van Mars willen oeffenen. Tot nut en gerief hun uit de braefste Autheuren vertolkt, door D.M. t'Amsterdam. [*Pyrotechnia, or artificial fireworks shown in running fire-breathing dragons and horses, castles, fighting ships, beacons, rockets, globes in the air, beaming stars ... and also the making of fire-balls, grenades, storm-wheels, stick-pots, petards ... useful for all young Constables & other amateurs who want to use them in the art of Mars ... For use and convenience translated from the best authors by D.M. of Amsterdam.*]

M060.3
(Pt.I) Baltes Boekholt: Amsterdam 1673
EngTP+TP+iv+192+22Pls.

M060.4
(Pt.II) Hugo Rijckhals: Rotterdam 1671
TP+152(-2)+viii+6Pls. NM — BL —

Pyrotechnia, of Meer dan honderderleye Konstvermakelijcks Vuurwerken. [*Pyrotechnia, or more than a hundred recreational artificial fireworks.*]

M060.5
Deerde en laatste Druck. Baltes Boekholt: Amsterdam 1678
EngTP+ii+170+iv+22Pls. 17cms. NM 0174936 BL —

It would seem that Part II was published first, in Rotterdam in 1671, and that Part I was first published in 1672. A second and third edition of Part I was then published in Amsterdam in 1673 and 1678. Sotheran records some details of all three editions. The plates are numbered consecutively from 1 to 29, but 24 is not called for in the list of plates at the beginning, numbers 1 - 22 being in Part I and 23 - 29 in Part II. It is clear from the text and the plates that BABINGTON's *Pyrotechnia* was the main source of this work. Nearly all BABINGTON'S plates depicting automata and dragons are here reproduced.

References: Berlin 3299 (1678). Haan 2970/1 & 3001a. CP (1673/71). RDML (1672/71)

MANUALE

Manuale del Pirotecnico: ricordi e memorie di un dilettante. [*Manual of pyrotechnics: memories and recollections of an amateur.*]

M070.1
Bianchi: Oderzo 1878
50p 16mo. NM — BL —

Manuale del Pirotecnico o l'arte di fare fuochi artificiale. [*Manual of pyrotechnics or the art of making fireworks.*]

M070.2
Croci: Milano 1886
24mo. NM — BL —

References: Pagliaini

MATYUKEBICH, F. F

Sobraniye formul i retseptov sostavam poteshnoy pirotekhnii. [*A collection of formulae and recipes for recreational pyrotechnic compounds.*]

M080.1
St. Petersburg 1861
NM — BL —

References: SHIDLOVSKY's bibliography

MAYER, Wolffgang Augustin

(Pt. I.) Lust- Lufft- und Feuer Kunst. Auss welcher ohne sondern costen und Mühe zuerlangen ... (Pt. II.) Kurtzer hoch gründiche Unterricht wessen sich ein jeder Constabel öder Büchsenmeister ... durch Marx Eysenträmer. [*The art of fireworks.*]

M090.1
Mattheus Schultes: Ulm M.DC.LXXX (1680)
TP(BR)+ii+17Pls(38Figs)+38+TP+41-78. 18cms.
NM — BL —

References: CP

McLAIN, Joseph Howard (1916-1981)

Pyrotechnics from the viewpoint of solid state chemistry.

M100.1
Franklin Institute Press: (Philadelphia) 1980
TP+xi+243illus. 24cms. NM — BL —

Unenthusiastically reviewed in *Pyrotechnics* No.VII May 1981. 'The book's preface is marked by its eloquent lamentation of the sparsity of pyrotechnic literature. ... this book is markedly sparse in its citations of literature devoted specifically to pyrotechnics. A number of accessible sources of information seem to be unknown to our learned author ... The chapter on "Fireworks" is a disappointing treatment both because of its brevity and because of the nature of its content ... Only nine formulae are quoted, and of these, four appear to be cribbed from previous publications without credit!'

References: Du Pont

MENESTRIER, Claude François

Advis necessaire pour la conduite des Feux d'Artifice ... [*Advice necessary for the management of fireworks.*]

M110.1
Lyon 1660
NM — BL —

He was responsible for a large number of designs and descriptions of fetes, celebrations and public events in the 17th century.

References: Lotz.

MEYER, Ernst Sigismund Christian von (1847-1916)

Die Explosivkörper und die Feuerwerkerei. Bearb. von Dr. E. v. Meyer ... [*Explosives and firworks.*]

M120.1
F. Bieweg und Sohn: Braunschweig 1874
viii+126. 22cms. NM 0525110 BL ML.b.79(2)

Part of Bolley's *Handbuch der Chemischen Technologie* No.23. (Bd.VI. 3 gruppe, 2 abth.). See also J. UPMANN's *Das Schiesspulver* ... which formed the first part of Volume 6 Section 2

References: Du Pont. CP (1874)

MEYER, Franz Sales (1849- ?)

Die Feuerwerkerei als Liebhaberkunst. [*The making of fireworks as an art for amateurs.*]

M130.1
von Seeman: Leipzig 1898
270+140illus. 23cms. NM — BL D

References: Du Pont. Keyser. G&O (1898)

MEYER, Capt. Franz Eduard Mori(t)z (1798-1838)

Die Feuerwerkerei in iher Anwendung auf Kunst, Wissenschaft und Gewerbe. Nach neuen Versuchen Bearbeitet von Dr. Moritz Meyer ... [*The making of fireworks as applied to art, science and industry. According to new experiments edited by Dr. M. Meyer*]

M140.1
Neue bear. J.A. Barth: Leipzig 1833
54p. 21cms. NM — BL 8715.c.45

Pyrotechnie raisonée, ou Application de la chimie aux artifices de guerre, ... traduite de l'allemand par T.Hippert. [*Descriptive pyrotechnics, or the application of chemistry to military fireworks.*]

M140.2
M. Hayez: Brussels 1836
ii+154. 21cms. NM — BL —

Lehrbuch der Pyrotechnik. ... Vollendetes Manuscript aus seinem Nachlasse. Herausgegeben und mit einem Nachtrage versehen von C. Hoffmann ...

M140.3
C. Heymann: Berlin 1840
viii+231+FPls. 21cms. NM 0531182 BL 8715.c.6

Traité de pyrotechnie, ... edite et augmente d'un appendice par C. Hoffman ... Traduite de l'allemand et augmente de notes par J.B.C.F. Neuens ...

M140.4
A. Leroux fils: Liege 1844
xviii+225+FPl. 27cms. NM — BL —

Michael Swisher say of these works that 'they are the first to bring a scientific and quantitive approach to the formulation of pyrotechnic compositions. They lay more stress on the *method* of working out a composition than on publishing large numbers of mixtures like CHERTIER, TESSIER and others, who present you with their *results* but do not discuss the method of arriving at these results.' Kayser lists a number of other works which deal with military and artillery subjects

References: Brock (1836). Kayser

MINIATUR-BIBLIOTHEK

Die Feuerwerkerei zur belustigung. [*The making of fireworks for recreation.*]

M150.1
A.D. Paul: Leipzig 1902
128(5illus). NM — BL —

A large number of mostly anonymous works are listed under this general heading of MINIATUR-BIBLIOTHEK in Kayser. I have not been able to find any further information.

References: Kayser

MITRA, Haridas

The Fireworks and Fire Festivals in Ancient India.

M160.1
Abhendananda Academy of Culture: Calcutta 1963
46p. 21cms. RLM 129075 BL (IOL) T.20701

MONTUCLA, Jean Etienne. (1725-1799) (Pseud. M. de C.G.F. See note under **OZANAM**)

MOORE, Dennis Times. See **PRACTICUS**

MOREL, A. M. Thomas

Traité Practique des Feux d'Artifice pour le spectacle et pour la guerre, avec les petits Feux de table, et l'Artifice à l'usage des Théâtres. [*Practical treatise of fireworks for celebration and war, with indoor fireworks and items for use in the theatre.*]

M170.1
Firmin Didot: Paris AN IX - 1800
½T+TP+164+11FPls+1FT. 20cms. NM 0767299 BL —

M170.2
Firmin Didot: Paris AN 1818
½T+TP+164+11FPla+1FT. 22cms. NM 0767300 BL —

The 1818 edition is identical to that of 1800, but it has been reset.

References: Du Pont. Brock (c.1812). BNC. CP (1800, 1818)

MORETTI, Tommaso (? -1675)

A General Treatise of Artillery: or, Great ordnance writ in Italian by Tomaso Moretti of Brescia ... Translated into English, with notes thereupon, and some additions out of French for sea-gunners, by Sir Jonas Moore, Kt. With an appendix of Artificial Fireworks for war and delight; by Sir Abraham Dager ...

M180.1
O. Blagrave: London 1683
TP+viii+Tab+124+74Pls. 18cms. NM 0772576 BL 534.c.21

The original 'Trattato dell Artigieria' was published in Venice in 1665.

References: Du Pont

MORTIMER, G. W

A Manual of Pyrotechny; or, A Familiar System of Recreative Fire-works.

M190.1
Simpkin, Marshall and Co: London MDCCCXXIV (1824)
vi+172+1FPl. 20cms. NM — BL 1034.k.38

Pyrotechny; or, A familiar system of recreative Fire-works.

M190.2
2nd ed.rev. J.S. Hodson: London [1853]
TP+xiv+126illus. 19cms. NM 0802616 BL 1034.k.39

M190.3
3rd ed.rev. J.S. Hodson: London [1856]
TP+xvi+131illus. 18.5cms. NM — BL —

The Pyrotechnist's Companion: Familiar system of Recreative Fire-works.

M190.4
Philadelphia 1845
NM 0802614 BL —

The Pyrotechnist's Companion: or, A Familar System of Recreative Fireworks. First American from the second London revised edition.

M190.5
H.C. Baird: Philadelphia 1852
½T+TP+viii+168(-12)illus.+24ads. 18cms NM 0802615 BL —

Pyrotechny or Recreative Fireworks.

M190.6
Paris (in English) 1852
NM — BL —

There is another issue of the first edition which omits the date 'MDCCCXXIV'. The preface to the 'Second edition' is still dated 1824. The 1852 American edition is virtually identical to the '2nd. London revised edition', this therefore slightly pre-dates the English version which is first advertised in the November 1853 issue of *The Publishers Circular*. The accompanying advertisement, quoting 'The Leader', says, 'A popular and familiar exposition of an art dreaded by mothers, gloried in by small boys and admired in its results by gaping thousands every summer night.' The third edition is advertised in the November 1856 issue of *The Publishers Circular*.

References: Du Pont. Brock (1824,1852). Berlin 3310 (1824). CP (1824, 1852, 1853)

MOUREY, Gabriel

Le Livre des Fêtes Françaises. [*The book of French fetes.*]

M200.1
Librairie de France: Paris 1930
½T+ColFP+TP+xii+380+xviii+306ill+11colPls. 28cms.
NM 0832870 BL 10167.s.1

Superb collection of plates and illustrations dealing with French fetes from the 15th to the 20th century. A large number of these depict the firework displays that were, inevitably, connected with these events. The earliest print that is reproduced is of a Naumachie of 1550, 'Il y eut eu un terrible feu de joye, veu le grand nombre et quantitie de feez artificielz qu'on avoit mis dedans'.

References: CP

MYDORGE, Claude (1585-1647) See D267/227

Recreation Mathematique ... augmentee de plusieurs beaux problemes par DHPEM.

M210.1
(4th ed) Rolet Bovtonne: Paris 1627

[Main TP] Examen dv Livre des Recreations Mathematiqves [Part III TP] Troisieme Partie des Recreations Mathematiqves, composee d'un Receuil de plusieurs plaisantes & recreatiues inuentions de feux d'artifice plvs la maniere de faire toutes sortes de fuzees, tant simples que double, auec leur composition le tout representé par figures. (Part IV TP) Nottes svr les Recreations Mathematiqves: ... par D.H.P.E.M. [Denis HENRION]. [*An examination of the book of mathematical recreations. (Part III). Third part of Mathematical recreations, containing a collection of many pleasant and recreative inventions of fireworks including the manner of making all kinds of rockets, single and double, with their composition all represented by figures.*]

M210.2
Rolet Bovtonne: Paris M.DC.XXX. (1630). (Part IV) Anthoine Robinet: Paris MDCXXX. (1630)
(I) TP+vi+8+280 (II) TP+ii+64(-4) (III) TP+ii+(68-106)+xii
(IV) TP+38. NM 0920701 BL 1651/593

M210.3

Paris 1634 NM — BL —

M210.4

Rolet Bovtonne: Paris M.DC.XXX.VIII. (1638). (Part IV) Rolet Bovtonne: Paris M.DC.XXXVIII. (1638)
(I) TP+vi+8+280 (II) TP+ii+64(-4) (III) TP+II+(68-106)+xii
(IV) TP+38. NL 0303587 BL 529.b.7

M210.5

A. Robinet: Paris 1639. NL 0303588 BL —

M210.6

Charles Osmont: Roven 1643. NL 0303592 BL —

M210.7

Rouen 1649. NL — BL —

Les Recreations Mathematiques avec l'examen de ses problemes ... Premierement reueu par D. Henrion. Depuis par M. Mydorge et tout nouuellement corrige & augmente. Cinquiseme & derniere edition. [*Mathematical Recreations with an examination of its problems, originally reviewed by D. Henrion. Then by M. Mydorge and all newly corrected and augmented.*]

M210.8

Cardin Besongne: Paris 1659
TP+xiv+416. 17cms. NL — BL —

M210.9

Cardin Besongne: Paris 1660. NL 0303627 BL —

M210.10

Cardin Besongne: Paris 1661. NL 0303628 BL 1607/494

M210.11

Vefue d'A Ferand: Rouen 1663? NL 0303629 BL —

M210.12

(6th ed.) J. Lucas: Rouen 1664. NL — BL —

M210.13

Francois Vaultier: Rouen 1669. NL — BL —

M210.14

C. Mallasis: Rouen 1669. NL 0303630 BL —

M210.15

David Berthelin: Rouen 1669. NL — BL —

M210.16
Deville: Rouen 1680. NL — BL —

M210.17
Jean Boulley: Rouen 1743. NL — BL —

A very confused situation! The main source of information for all these editions is Backer, but in some cases his listings are far from clear. For example the BL copy of the Besongne edition of 1661 has a title-page date giving 'MDCLXI' but Backer quotes this 'cinquième et dernière édition' as 'Paris, Cardin Besongne 1659 (c'est 1660). Dans cette édition les notes de Denis Henrion sont distribuées, dans la première partie, avec chaque remarque de Cl. Mydorge ... Il existe plusieurs autres éditions de ces Recreations, soit avec les notes, soit sans notes. La dernière, peut-être entirement anonyme, est de Lyon, Deville 1680. (Barbier 15414).' A few columns later he adds a Rouen 1743 edition.

The original 1628 edition of *Recreations Mathematiques*, the first to contain the Firework section, included 12 small woodcuts, some or all of these are recopied in the other commentary editions and translations. The first time that D.H.S.E.M's (Henrion) notes appear are in the Boutonne edition of 1627, but they do not contain the firework section. (See note to ETTEN). The 1630 and 1638 Mydorge editions have all 12 woodcuts, the 1661 edition has 13 because the first one is reproduced twice. In addition the 1661 edition is continuously paged without the usual additional sectional title-pages. These are the only ones that I have been able to examine.

The NUC confuses and duplicates some editions by listing them variously under 'Cyriaque de Mangin' (for Henrion) and 'Mydorge'. Most of those quoted above are under 'Leurechon'.

References: Backer (1630, 1634, 1638, 1639, 1649, 1659, 1660, 1663, 1664, 1669 (Vaultier & Berthelin), 1680, 1743). BNC (1630, 1639, 1659, 1660) CP (1630)

NÆSER, A

Lystfyrvaerkeren. Kort Begred over Lystfyrvaekerier. [*Fireworks. A short description of recreative fireworks.*]

N010.1
H.P. Muller: Copenhagen 1854
NN — BL —

See also BEGRED.

References: VAHL. Barfod

NAMBO, Heizo (1892- ?)

Ryokoku hanabi no genso. [*The origins of Firework Displays on the Ryokoku Bridge.*]

N020.1
1972

NN — BL —

References: HOSOYA bibliography

NAN, Lou

Huo yao han huo yao wuch'l.

N030.1

NN — BL —

Juvenile Firework literature

References: NUC SC 1965-69

NAT, E. A

Praktika dlya pirotekhnokov. [*Practical handbook for pyrotechnists.*]

N040.1
St. Petersburg 1845

NN — BL —

References: SHIDLOVSKY's biblioggraphy

NEU-EROFNETE

Die neu-eröfnete Arsenal, worinnen der galanten Jugend und andern Curieusen, insonderheit aber den Reisenden das Merckwürdigste von der Artillerie kürtzlich und solcher Gestalt adgehandlet wird; das ein jeder von Canoniren und Bombardiren, von Ernst- und Lust-Feuern ... [*The newly-inaugurated arsenal, in which the most noteworthy aspects of artillery and the like are dealt with.*]

N050.1
Benjamin Schillern: Hamburg 1702
143+Pls. 16cms. NN 0131148 BL 7845.a.4

N050.2
Benjamin Schillern: Hamburg 1704
TP+144(-2)+Pls. 16cms. NN — BL 1506/782b/2

NEW

A New and Comprehensive Edition of the Art of Making and Managing Fireworks. See **ART**

NIDA, C.A. von

Katechismus der Lustfeuerwerkerei. Kurzer lehrgang für die Gründliche Ausbildung in allen Teilen der Pyrotechnik. [*Catechism of firework making. A short course for basic training in all aspects of pyrotechnics.*]

N060.1
J.J. Weber: Leipzig 1883
viii+214p. 16mo. NN 0254711 BL —

References: G&O. NYPL

NIKOLAYEV, N

Pirotekhnik-lyubitel'. [*Amateur pyrotechnist.*]

N060.1
1899
NN — BL —

References: SHIDLOVSKY's biliography

NISHIZAWA, Yushichi (1883-1943)

Nippon kajutsu-ko. [*Examination of Japanese Pyrotechnics.*]

N070.1
Shuhokaku: Tokyo 1927
384p. 20cms. NN — BL —

Nippon kajutsu-ko yakuho no maki. [*Pyrotechnic compositions in Japan.*]

N070.2
Togaku-sha: Tokyo 1935
619p. 22cms. NN — BL —

Hanabi no kenkyu. [*Investigative study of Fireworks.*]

N070.3
Uchida rokakuho: Tokyo 1938
710p. 23cms. NN — BL —

Shinheiki kagaku hanabi no kenkyu.

N070.4

240+26Pls. 23cms. NN — BL —

SHIMIZU's note on these books says that, 'they are similar in content containing the various types of pyrotechnic items, history, technique, customs, etc., collected in a disorderly fashion. They are of more use to the hobbyist than to the practical manufacturer'.

References: NUC SC1975 (Shinheiki)

NORTON, Robert (? -1635)

The Gvnner, showing the vvhole practise of Artillerie: vvith all the appurtenances thereunto belonging. Together with the making of extraordinary artificiall Fire-workes, as well for pleasure and triumphes, as for warres and seruice ... [*Slight variations in long-title.*]

N080.1
A.M. for H. Robinson: London 1628
EngTP+xii+158+ii+30Pls. NN 0326758 BL ML.p.11(2)+

N080.2
H. Robinson: London 1664
EngTP+xii+158+ii+30Pls. NN — BL 534.k.33

N080.3
(Reprint)Da Capo Press. Theatrvm Orbis Terrarvm: Amsterdam/New York: 1973
(½T+TP)+EngTP+xii+158+ii+30Pls. 30cms. NN — BL 1653/617

The Gunner's Dialogve with the Art of Great Artillery.

N080.4
Thomas Tap: London 1628
NN 0326760 BL —

N080.5
np. : London 1643
TP+ii+(71-92)+FT+2FPls. 18cms. NN — BL 57.B.25(3)

HODGKIN comments on *The Gunner*; 'Pages 149-158 are devoted to Artificial Fireworks "for Tryumph and Seruice". Of the favourite device of "Rockets that runne on lynes" Norton says "they ought to be very carefully made whether they are to be double or single, or those that carie Draggons, Men, Shipps, or other shapes in motion, least they shame their Master." '

The Gunner's Dialogue contains fifty 'Questions & Answers' dealing with the matters that a Gunner should be familiar with. The last four questions deal with 'Fire-workes of Triumph, for pleasure'. The 1643 edition is paged continuously, following BOURNE's 'The Art of Shooting in Great Ordnance ...'. Norton's 'The Art of Great Artillery' (STC18676) is a separate work that does not deal with recreative Fireworks.

References: Du Pont. Cockle 114. Brock (1628, 1643). CP (1973). STC18673, 18674

NYE, Nathaniel (1624- ?)

The art of Gunnery ... To make divers sorts of Artificial Fireworks ...

N090.1
William Leake: London 1647
102+2Pls. 16cms. NN 0361966 BL 1607/962

N090.2
William Leake: London 1648
102+2Pls. 16cms. NN — BL 1398.a.22

N090.3
William Leake: London 1670
102+Pls. 16cms. NN 0361967 BL 1651/848

Sotheran says: 'This is one of the rarest English works on Gunnery. The author ... also claims in the Address to the reader that some of his Fireworks are quite new and states, "I do protest to thee, that I have made and still do make practice of them myself, having by experience to have found them to be the best of all others that I have read of or that are taught by Bate, Babington, Norton, Tartaglia or Malthus."

HODGKIN says of it, 'A rare little book (largely plagiarised from Malthus) both as regards text and engravings, that which I here reproduce being actually printed from the same copper which was used in the edition of that author of 1629. ... Nye acknowledges his obligations in a very half-hearted fashion. He says:- "I will not teach how to make Moulds, Needles, &c. for such men as are curious in these things, let them buy Master Bate, Master Babington, or for a need Master Malthus Fire-works."

References: Du Pont. Brock. Wing N.1481-3. NYPL (1647)

OGATSU, Kyosuke (1919-   )

Nippon hanabi-ko. [*Investigation of Japanese fireworks.*]

O010.1
Mainichi Newspaper Office: Tokyo 1979
206p. 20cms. NO — BL —

Hanabi o ageru. [*The firing of fireworks.*]

O010.2
Popular Books: Tokyo 1981
221p. 20cms. NO — BL —

The author is the director of the famous Japanese firm of Marutamaya-Ogatsu. 'A book for everyone' is SHIMIZU's summary of the 1979 edition. It has chapters on the Edo (1603-1868) makers and festivals as well as many other famous firework makers and celebrations. The 1981 edition is similar to the earlier one but specifically published as a children's book.

References: CP (1981)

OMODEI, Francesco

Dei razzi osservazioni storiche del cavaliere Francesco Omodei [*Observations of the history of rockets.*]

O020.1
Stamperia Reale: Torino MDCCCXXXV. (1835)
TP+32(-2). 27cms. NO — BL —

References: CP

ORVAL. See PERRINET D'Orval

OSSORIO, D. Marcello Cala

Istituzioni di Pirotecnia. Per istruzione di coloro che vogliono apprendere a lavorare i Fochi d'Artifizio. [*Institution of pyrotechnics. For instructing those who wish to work in artificial fireworks.*]

O030.1
Stampera Reale: Napoli 1819
TP+iv+236+ii+TP+23Pls+FT. 29cms. NI 0176761 BL —

This seems to cause cataloguing difficulties both to booksellers and the NUC. The author is clearly named in the dedication at the end of the book as 'D. Marcello Cala Ossorio di Villanova' and the monogram on the Title page has 'Marcello Cala Ossorio' inscribed inside it. A comprehensive work containing much detailed description of Italian firework practice in the early 19th century including some very elaborate set-pieces. I possess a second copy of this work which has 19 additional pages of manuscript text as well as two additional hand-drawn plates tipped in.

References: Du Pont. CP

OUD-SOLDAAT

Handleiding tot het vervaardigen van Lustvuurverken, voor oud en jong door een oud-Soldaat. [*Treatise for the manufacture of recreative fireworks, for young and old by an ex-soldier.*]

O040.1
C. Schook: Gorinchem 1873
NO — BL —

References: Brinkman

OUGHTRED, William See ETTEN, Henri van

OZANAM, Jacques (1640-1717)

Récréations Mathématiques et Physiques. Qui contiennent ... de pyrotechnie ... [2 *Vols. or 2 Vols. in 1.*]

O050.1-4
G. Gallet: Amsterdam 1693, 1696, 1698 & 1700

O050.5-7

J. Jombert: Paris 1694, 1696, 1697.	NO 0189051 et seq.	BL 8531.aaa.23 (Paris 1696)

Récréations Mathématiques et Physiques, Qui contiennent ... de pyrotechnie ... Nouvelle edition revue, corrigée & augmentée. [4 *Vols.*]

O050.8-13
C.A. Jombert: Paris 1723, 1725, 1735/36, 1735/41, 1750, 1770

O050.14

J.Rollin: Paris 1750 Vol.I TP+xiv+460+xx+32Pls Vol.II TP+ii+462+xiv+55Pls Vol.III TP+482+xii+30Pls Vol.IV TP+viii+446+iv+17Pls. 17cms.	NO 0189068 et seq.	BL 8534.bbb.29 (1741) and others

Récréations Mathématique et Physiques, Par feu M. Ozanam ... Nouvelle édition, totalement refondee a considérablement augmentée par M. de C.G.F. (Chanla, geometre, forezien. pseud. of MONTUCLA: J.E.). Tome troisieme continent ... la Pyrotechnie. [4 *Vols.*]

O050.15

C.A. Jombert: Paris M.DCC.LXXVIII (1778).	NO 0189088	BL 1474.bb.22

Récréations Mathematique ... par M. de M . [*i.e. MONTUCLA*]

O050.16

F. Didot: Paris 1790.	NO 0189093	BL 8530.bbb.14

Recreations Mathematical and Physical; Laying down, and Solving Many Profitable and Delightful Problems of, Arithmetick, Geometry, Opticks, Gnomonicks, Cosmography, Mechanicks, Physicks, and Pyrotechny. By Monsieur Ozanam Professor of the Mathematiks at Paris. Done into English, and illustrated with very Many Cuts.

O050.17
R. Bonwick ... and Ralph Smith: London 1708
TP+xxxvi+530inc28Pls+iv. 19cms. NO 0189050 BL 8534.bb.2

Recreations for Gentlemen and Ladies, or, Ingenious Amusements, being curious and diverting sports and pastimes, natural and artificial, with many inventions, pleasant tricks on the cards and dice, experiments, artificial fireworks, and other curiosities, affording variety of entertainment. Written in French. Translated into English. [*Slight variations in long-titles.*]

O050.18
James Hoey: Dublin 1756
155+ix. 16cms. NO 0189042 BL —

O050.19
James Hoey: Dublin 1759
192p. NO — BL —

O050.20
Peter Hoey: Dublin 1790
226+xii. 16cms. NO 0189043 BL D

The French editions of 1778 and 1790 were revised by Montucla and were subsequently translated by HUTTON and became *Recreations in Mathematics and Natural Philosophy*. The NUC lists all the HUTTON editions under OZANAM.

The only clue to the author of the first English edition of 1708 is given in the 'To the Reader' where it states that, 'the Care of Translating being committed to a Gentleman of great Ingenuity, and well-vers'd in these Sciences; who had seen but a few Sheets from the Press, when he was snatched from hence by untimely Death.' The last three plates in the Pyrotechny section are misnumbered 24-26 instead of 26-28. Furthermore, some of the figure numbers are duplicated; an error that also occurs in the French versions.

The figures on the 3 plates dealing with Fireworks in the English translations are identical to those in OZANAM but have been re-engraved. The 38 'Problems of Pyrotechny' occupy pages 481-530 and although not a literal translation, the titles of the 'Problems' are the same as those in OZANAM. (At least so far as the 1770 edition is concerned; the only one I have seen). They deal with gunpowder and its constituents, various types of rocket, serpents, lances, stars, rain, sparks, fire-balls for use in water and in the air (shells), wheels, fire-clubs, and a few items for 'Service in War'. Full of good practical advice and detail. See also under HUTTON.

References: BNC. SOT (1708). Hall 212-4 & 6. Stott 518-20. CP (1770, 1708 Eng.)

OZERKOV, D

Obshchedostupnaya pirotekhniya. [*Pyrotechny for everyone.*]

O060.1
1897
NO — BL —

References: SHIDLOVSKY's bibliography

P, G.F See PARDON, George Frederick

PACINI, V

Il Dilettante de Pirotecnia, ovvero regole da tenersi per lavorare i fuochi d'artificio e colorati. [*The amateur pyrotechnist, or rules to be observed when working with fireworks and colours.*]

P010.1
Campolmi: Firenze 1845
64p+FPl. 12mo. NP — BL —

References: BC

[PARDON, George Frederick] (1824-1884)

Games for All Seasons: consisting of In-door and Out-door Sports ... Magic and Legerdermain, Fireworks etc. A Sequel to 'Parlour Pastimes'.

P020.1
James Blackwood: London 1858
ix+208(-10) illus. 17cms. NP 0078455 BL D

The BL Catalogue lists another edition of c.1868; both were destroyed.

References: Stott 311

PARTINGTON, James Riddick

A history of Greek Fire and Gunpowder.

P030.1
W. Heffer & Sons: Cambridge (1960)
ii+illus+TP+xii+381+21figs. 26cms. RLP 024544 BL 8701.k.22

A classic and very well detailed and documented work on the subject. Has quite a large amount of material concerning Chinese fireworks quoted from Jesuit missionaries such as AMIOT and INCARVILLE.

References: CP

PERETSDORF, Baron M. J. D. Ravichio de (1767-1844)

Traité de Pyrotechnie Militaire, comprenant tous les artifices de guerre en usage en Autriche. Traduit ... par J.Ravichio de Peretsdorf.

P040.1
F.G. Levrault: Paris/Strasburg ... 1824
TP+iv+264+36Pls. NT 0307240 BL ML.i.52

The NUC listing is under 'Traité'. Many of the devices that are described; rockets, match, signal flares, stars, maroons etc., could just as easily be used for recreative purposes. Numerous plates together with manufacturing proceses described in detail.

References: Brock

PERFECTION

The Perfection of Military Discipline After the Newest Method, as Practised in ENGLAND and IRELAND, ... To which is added, as a Second Part, The Art of Gunnery; or, The Compleat Engineer, Shewing the Practice of the Ordnance, Mortars, etc. with the manner of making and using of Fire-works for War, at Sea and Land ...

P050.1
N. Boddington: London 1690
FPl+TP+ii+156+FPl. 17cms. NP 0236838 BL 8824.a.27

P050.2
2nd ed. N. Boddington: London 1691
NP — BL —

P050.3
4th ed. corr. N. Boddington: London 1702
EngPl+TP+ii+170+FPl. 17cms. NP — BL 1140.b.6.+

The 1690 edition does not contain any material on the use of fireworks other than for war. However the 1702 edition has an additional chapter heading on page 153 entitled, 'The Second Part ... being the Art of Gunnery and Making Fire-works for War, Recreation etc. and other Matters ...'. Despite its lengthy title only the last two pages are devoted to a 'Treatise of Artificial Fire-works' in which a few words concerning rockets are found. In the BL catalogue it is listed rather confusingly under 'England', althought the new ESTC has it under 'Perfection'.

References: WING P1532 (1690) & P1533 (1691). Arber (1690, 1691). RAI (1690).

PERIODICALS See BMAGAZINES

PERRINET d'ORVAL, Jean Charles (1707-1780)

Essay(i) sur les Feux d'Artifice pour le spectacle et pour la guerre. [*Essay on fireworks for celebrations and war.*]

Q477/$150 15R58
S6/3/78/120/£170
ABPC 78 $275

P060.1
Coustelier: Paris 1745. M.DCC.XLV. (1745)
TP+xiv+224+13Pls. 20cms. NP 0250650 BL 535.d.1

Traité des Feux d'Artifice pour le Spectacle et pour la guerre.

P060.2
Wagner et Muller: Berne 1750
TP+vi+242+15Pls NP — BL —

Manuel de l'artificier.

P060.3
Editeurs du Journal Helvetique: Neufchatel MDCCLV (1755)
TP+iv+128+ii(corriger)+11Pls. NP — BL —

Manuel de l'artificier. Séconde édition, révué, corrigée & augmentée.

P060.4
Charles-Antoine Jombert: Paris M.DCC.LVII. (1757)
½T+TP+iv+141+iii+12FPls+5FT. 17cms. NP — BL 1568/219

Perrinet d'Orval was an officer in the French army Corps of Engineers. Due to errors made by earlier bibliographers, some of his works were credited to FREZIER, a contemporary pyrotechnist. These errors sometimes continue to this day, and even the NUC seems somewhat confused about the matter. All the above listed works are clearly credited to Perrinet d'Orval in Vincent Poupard's *Histoire de la Ville de Sancerre* (2nd ed. 1838 BL 10170.ccc.18). Under the listing of*Manual de l'Artificier* he says, 'seconde édition revue et augmentée imprimee a Paris Chez Jombert'. He goes on to say, 'Il est aussi auteur de plusieurs articles de pyrotechnie insertés dans le dictionnaire encyclopaedique', (i.e. Diderot et d'Alembert). Part of the confusion probably stems from the entry in J.M. Querard's *La France Litteraire* Paris 1845. Under the entry for Perrinet d'Orval's*Essai sur les Feux d'Artifices* he adds, 'Le même ouvrage, sous le titre:- *Traité des feux d'Artifice pour le spectacle et pour la guerre.* Nouvelle édition, toute changée et considerablement augm. Par M.F.F.(Frezier) D.D.F.D.B. Paris, Nyon Fils 1747.' This is clearly NOT the same work as Perrinet's *Essai*.Sotheran in his *Bibliotheca Chemico-Mathematica* lays the blame on Barbier's *IDictionnaire des ouvrages Anonymes* Paris 1875, which, in a note to FREZIER's *Traité* of 1747 states the work to be by Perrinet d'Orval. Confusion all round! FREZIER, in his introduction to the 1747 edition claims that it was as a result of Neaulme publishing, in 1741, a pirated edition of his own *Traité*, first published in 1706, and the recent appearance of Perrinet's *Essaii* that encouraged him to produce the revised version. Apart from any other consideration, the first two works of Perrinet both contain material relating to military uses of fireworks, a subject not dealt with by FREZIER. This confusion is not confined to matters of authorship. Poupard lists 14 plates for the *Essai* (which in some issues is spelt *Essay* on the title-page). Only 13 are mentioned in the text and there are only 13 in the versions I have seen. In the *Manuel* of 1757 there are references on pages 109 & 135 to 'Pl.13'. This plate number does not exist, but the references are clearly intended to refer to Plate 11.

The first edition of *Manuel de l'Artificier* published in 1755 is very scarce, only twelve copies are supposed to have been issued. The 1757 edition contains a large section dealing with Chinese fireworks as described by Pierre d'INCARVILLE. Many of the figures that appear on the plates of this edition are very similar to those that were used in the original *Essai*.

Michael Swisher has commented that, 'all these books are practical manuals that reflect the craft of the working pyrotechnist in the mid-18th century. Their flavour is perhaps best illustrated by this extract to the preface to the 1750 edition, "Of all the authors who have written in different languages on the art of making them [fireworks], I have seen none but Casimir Siemienowies (sic) who has treated them appropriately, and as a man who was well versed on the subject. The greater part of those who have followed him, have given nothing but truncated compilations, and badly understood ones, of the work of this Pole, who ought to be regarded as the master of pyrotechnists. Ozanam and many moderns, who like him, have wished to speak of this art, of which they have not the least notion, have printed nothing but absurdities. The art having been abandoned to purely mechanical hands, has not however gone without much improvement since the time of Siemienowies, solely by practical means and without the aid of writers, who have contributed practically nothing to it. It is an avowal I cannot prevent myself from making, although this observation would appear to destroy in some way the utility of instructional books, and consequently my own. For the rest, pyrotechny has not abounded in writers; M. Frezier and I are the only ones who for the past fifty years have undertaken to treat it methodically; and I have no one but him to emulate in France.'' I think this is not only a fair illustration of Perrinet's approach, but also a fair criticism of most of the pyrotechnic literature he discusses'. (See also the note to FREZIER).

References: Brock (1740[?]), 1745). Du Pont. Berlin 3302 (1745). CP (1745, 1757)

PETROV, M

Obshchedostupnaya pirotekhniya. [*Pyrotechny for everyone.*]

P070.1
Printing House Cooperative 1927
NP — BL —

References: SHIDLOVSKY's bibliography

PIRKER, Adolf

Die vornehme Kunstfeuerwerkerei. Wertvolle Fundgrube und verlasslicher, praktischer Rathgerber zur Anfertigung, Dillettanten und Berufs-Pyrotechniker ... [*The making of fireworks, a veritable mine of reliable, practical advice on manufacturing for amateur and professional pyrotechnicians.*]

P080.1
Ferd. v. Kleinmayr: Klagenfurt 1892
xi+178+4FPls. 22cms. NP — BL —

References: Du Pont. Kayser. G&O

PIROTHECNIE

Pirothecnie entretenida, curiosa y agradable de fuegos recreativos, con varias invenciones y secretos, y algunas ideas generales para que cada uno pueda formarse otras a su modo. Recogidas y ordenadas por un curioso. [*Entertaining pyrotechnics, curious and agreeable recreative fireworks with various inventions and secrets and some general rules by which anyone can make others of his own invention. Collected and edited by an inquisitive one.*]

P090.1
Viuda de Ibarra: Madrid MDCCXCIX (1799)
xiv+204+3Pls. 15cms. NP 0385341 BL —

P090.2
(Reprint) V. Vindel: Madrid 1955
NP — BL —

Palau y Dulcet *Manuel del Libero Hispano-Americano* gives PIROTECHNIE with a date of 1798

References: NYPL (1799). BC (1955)

PLATO, Alexander

Beschreibung einer neu erfundenen viel Holz und Zeit ensparenden Methode, den Salpeter zu sieben, nebst einer Anweisung zur Feuerwerkkunst. [*Description of a newly discovered method for making saltpeter which saves a great deal of wood and time, together with instruction on the art of fireworks.*]

P100.1
Christian Fr. Gutsch: Breslau 1792
TP+iv+116(-8)+3FPls. NP — BL —

P100.2
2e Aufl. F.L. Supprian: Leipzig/Breslau 1797
TP+iv+116(-8)+3FPls. NP — BL 1044.h.27/3

The first edition is very scarce as it is not listed by Kayser or other major reference works. It was however recently offered by an American bookseller. The book deals mainly with the manufacture of saltpetre and rockets, listing over 100 formulae of negligible difference.

References: Kayser 1797

PLIMPTON, George

Fireworks. (A History and Celebration).

P110.1
Doubleday & Company Inc.: New York 1984
½T+TP+iv+286(-8)illus. 26cms NP — BL —

The author, who is well known as an American writer and TV figure, gives a highly personal account of his infatuation and involvement with fireworks. In addition it includes an extensive glossary of firework terms and a description of many of the major firework events throughout the world. Although not strictly speaking a 'history', like BROCK's *A History of Fireworks,* it nevertheless contains over 100 well and copiously captioned colour and black-and-white photographs which cover an extremely wide range of pyrotechnic displays, ephemera and other items from all over the world from the 16th to the 20th century.

PRACTICAL hints on Pyrotechny. See **ELRINGTON, J.P**

PRACTICUS (Pseud. Dennis Times **MOORE**)

Papers on Pyrotechny. [Articles in *The Boy's Own Magazine.*]

P120.1
Ward, Lock & Tyler: London 1864-67

Vol.4. No.20 August 1864 p153
No.21 September 1864 p245
No.22 October 1864 p357
No.23 November 1864 p418
No.24 December 1864 p517

Vol.5. No.25 January 1865 p81
No.27 March 1865 p255
No.28 April 1865 p349
No.30 June 1865 p543

Vol.6. No.32 August 1865 p158
No.34 October 1865 p336
No.36 December 1865 p510

Vol.9. No.51 March 1867 p175
No.52 April 1867 p235 NM — BL PP5993d

Pyrotechny; or, the art of making Fireworks, at little cost, and with complete safety and cleanliness.

P120.2
Ward, Lock & Tyler: London c.1871
½T+TP+xvi+198+Cat+11Pls.(4F). 19cms. NP 0744800 BL —

A Manual of Pyrotechny. For the use of those who desire to prepare Fireworks for Exhibition.

P120.3
Bolton & Co./Wm.Bailey & Son: London/Wolverhampton 1872
TP+x+152illus. 22cms. NM — BL —

The articles that PRACTICUS wrote initially for the *The Boy's Own Magazine*, edited by S.O. Beeton, were subsequently republished in *Pyrotechny* unchanged except for a few words and tenses in the opening and closing paragraphs of each article. There were no articles printed in 1866 because, as PRACTICUS explains at the beginning of the article in No.51, he had been ill. At the end of the last article the publishers announced that they intended to issue them in book form, which they duly did. When bound the magazines are entitled *Boy's Own Volume*. Brock gives the date of *Pyrotechny ...* as 1865 which seems unlikely as the series of articles on which the book was based was not finished until 1867. Six of the original 14 articles were also reprinted in one of Beeton's Boy's Own Books, *Stirring Adventures Afloat and Ashore* edited by G.A. Henty and published by Ward, Lock: London (1892). The same six articles were again used in one of the series of Beeton's Boy's Own Magazine (Vol.VI) published by Ward, Lock: London 1900. *The American Catalog*, published in 1881 and listing books published up to 1876, lists *Pyrotechny; or, the art of making Fireworks* with no author given, but then published by Baird.

References: Brock (1865[?], c.1870). CP (1864, 1865, 1871, 1872)

PYROPHILO(US)

Pyrobolio succincta d.i. Kurtzbündige Feurwerckerey Kunst, worinnen die Ernst- und Lust-Feuer erkläret und in den beygefügten Kupfern vorgestellet. [*Brief description of the art of fireworks*]

P130.1
Georg Wilhelm Kuhnen : Frankfurt & Leipzig 1703
TP+ii+72+8Pls.(1F). NP — BL 8823.h.38/2

Forms the second part of the author's *Ars tormentaria enucleata - Buchsenmeisteren Kunst*. The plates depict, apart from the usual military pots, fire-arrows and lances, excellently drawn Saxons, wheels, rockets, pyramids, Roman candles and shells.

PYROTECHNICIAN, Juvenile See 'The **COMPLETE** Art of Making Fireworks'

PYROTECHNY

Pyrotechny, or the Art of Making Fireworks, containing remarks on what is or is not dangerous etc. Directions for making Rockets, Squibs, Crackers, Catherine Wheels ... with all the various coloured fires etc. etc.

P140.1
W.G. Kerton: London 1852
TP+18p. NP — BL 1398.a.3

The 9 leaves are numbered '37-54', so they presumably form part of another work

PYROTHECNIE See GRIGNON & LALAING

REILLY, Joseph (1889 - ?)

Explosives, Matches and Fireworks.

R010.1
Gurney and Jackson: London/Edinburgh 1938
½T+TP+viii+172. 25cms. NR 0144527 BL 8903.i.30

R010.2
D. van Nostrand: New York 1938
½T+TP+viii+172. NR 0144528 BL —

R010.3
(Reprint) Atlan Formularies: California 1977
NR — BL —

The reprint appears in *Poor Man's James Bond* edited by Kurt SAXON. (Pages 76-120). Deals in the main with the testing and analysis of materials used in the manufacture of matches and some fireworks. The formulae are out of date even for 1938.

References: CP (1938)

REINAUD, Joseph-Toussaint (1795-1867) and FAVE, General Ildephonse (1795-1867)

Histoire de l'Artillerie. 1 ere. partie. Du Feu Grégois des Feux de Guerre et des origines de la poudre a canon, d'après des textes nouveaux. [*History of Artillery. Greek fire and the origins of gunpowder. In two parts, Text and Plates*]

R020.1
J. Dumaine: Paris 1845
(Text)TP+288. NR 0147222 BL 1398.e.29

R020.2
(Plates)½T+TP+ii+17Pls.(13col). 27cms.
NR — BL 1398.k.4

The appendix to the text 'NOTE F' includes a 29-page reprint of INCARVILLE's description of 'les feux d'artifices Chinois'. See also note under AMIOT.

References: Brock. Du Pont. CP (Pls.)

RIFFAULT See BOTTEE de TOULMONT

RISHO (pseud.)

Hanabi hiden-shu. [*A Manual handed down secretly on how to make fireworks*]

R030.1
By the author: [Osaka] c.1825
39 leaves. 15x11cms. NR — BL —

The following note has been kindly provided by the British Library's Department of Oriental Manuscripts and Printed Books.

'This book gives detailed instructions of many different types of firework. It is divided into two parts. The first describes 26 kinds of miniature fireworks for domestic use, with names quite similar to those in current use in the West: 'Weeping Willow', 'Tiger's Tail', 'Thousand Fireflies'. The second part has more elaborate varieties to be used in public displays. For each piece Risho details the size, the exact quantities of the ingredients, and the method of assembly. In most cases he illustrates the piece itself, the crucial assembly stages, or the intended effect. The book opens with a very fine double-page woodcut, delicately coloured, showing a Firework Display by the Ryogokubash bridge in Edo (now Tokyo): such displays were held most years between 1733 and 1962 when they were banned by the city council because of the traffic jams that they caused. Fireworks were introduced into Japan by the Dutch traders in the 16th century, but they did not become commercially profitable for the Japanese manufacturers until the 1840s. Risho's delightful little book seems to be an early expression of the new possibilities. In his postscript the author announces an intended sequel, but that does not seem to have materialised. His manual itself is quite rare: the authoritative Kokusho somokuroku (the National Bibliography of books and manuscripts published in Japan before 1868) list no more than six copies extant in Japanese libraries. 39 leaves with one double page, 12 full-page and 34 smaller illustrations, nearly all coloured by hand. Bound in traditional fukurotoji (thread sewn 'bag-binding'); original blue paper covers.'

Even more significant is the fact that it appears to be the ONLY book, as opposed to manuscripts of which about a dozen are listed by the Kokusho, to have been published in Japan up to 1868 dealing with fireworks as it is the only one listed under titles beginning with the word 'Hanabi'. The formulae given in the book show that the only chemicals used in Japanese fireworks until 1825 were potassium nitrate, sulphur, hemp charcoal, iron dust and camphor. For a long time this book played a leading role in the production of fireworks in Japan. It is still practical today as it provides valuable information on the use of black powder type of compositions.

References: CP

ROMANUS, Adrianus (1561-1615)

Pyrotechnia hoc est, De Ignibvs Festivis, iocosis, artificialibvs et seriis, variisqve eorvm structuris. Libri dvo. [*Pyrotechnics, that is, concerning festive fire, joyous, artificial and serious, and their various structures. Two volumes.*]

R040.1
In Officina Paltheniana: (Francofurti) M.DC.Xl. (1611)
TP+ii+71(26WC). 18cms. NR 0399158 BL 534.f.38/1+

Sotheran says: 'A very rare and curious work which was unknown to the author's bibliographer in NBG, where it was stated that the author was a native of Louvain'. Hodgkin considered that, 'this very unpretentious volume is of great value in the history of the pyrotechnic art as it gives the earliest systematic classification of pieces and enables us to see at a glance what materials the arranger of a firework display had at his command at the beginning of the seventeenth century.'

Michael Swisher has seen a copy with a 'Herbipoli' (Wurzburg) origin where Romanus was Professor of the Faculty of Medicine.

References: Cockle 935. Du Pont. Hodgkin. CP

ROSSIGNON, Julio

Manual del Cohetero y Polvorista. Compendio de Pirotechnia. (Enciclopedia Popular-Mejicana). [*Manual of rockets and gunpowder manufacture.*]

R050.1
Rosa y Bouret: Paris 1869
15cms. NR — BL —

References: MS

RUGGIERI, Claude Fortuné

Elémens de Pyrotechnie, divisés en cinq parties; ... avec un vocabulaire des termes compris dans cet ouvrage. [*Elements of Pyrotechnie, divided into five parts, with a vocabulary of the terms used in this work.*]

R060.1
Barba ... : Paris an dix 1802 HAY CINTRA
xxiv+390+24Pls. 21cms. NR 0505071 BL — 8/88 £375

R060.2
Barba ... : Paris 1810
21cms. NR — BL —

Elémens de Pyrotechnie, divisés en cinq parties; ... Suivis d'un Vocabulaire et de la Description des quelques Feux d'artifice, etc. Seconde édition, Révué corrigée et augmentée de nouvelles Découvertes et Inventions faites par l'Auteur; telles que le Feu Vert pour Palmier, les Baguettes détonnantes pour Fusées volantes, etc.

R060.3
Barba; Magimel: Paris 1811
½T+TP+xl+432+27FPls. 21cms. NR — BL 8908.aaa.12

Élémens de Pyrotechnie ... des principaux Feux d'artifice qui one été tirés à Paris depuis 30 ans, etc. Troisième Édition, Revue, corrigée et augmentée de quatre articles et deux planches pour les nouvelles pièces ...

R060.4
Bachelier: Paris 1821
½T+TP+xxii+462+30Pls.(29F). 21cms. NR 0505072 BL —

Pyrotechnie Militaire ou traité complet des feux de guerre et des bouches a feu.

R060.5
Barba, L'Auteur, Magimal, Patris: Paris 1812
½T+TP+xxiv+368(-20)+8FPls. 20cms. NR 0505075 BL 8826.ff.29

Précis Historique sur les Fêtes, les Spectacles et les réjouissances publiques. [*Concise history of fetes, spectacles and public rejoicings.*]

R060.6
l'Auteur, Bachelier, Delauney, Barba: Paris 1830
½T+TP+viii+380. 21cms. NR 0505074 BL 583.f.31

GERMAN EDITIONS

Praktischer Unterricht in der Feuerwerkerkunst für Dillettanten und angehende Feuerwerker. Oder, Vollstandige Anweisung, alle Arten von Land-, Wasser-, und Lustfeuer-werken, so wie auch Feuerwerke zu Luftballons ... Nach Ruggieri bearbeitet von Fr. J.Hartmann ... [*Practical instructions on the art of fireworks for amateurs and fireworkers.*]

R060.7
G. Basse: Quedlinburg unn Leipzig 1832
xii+178+29FPls. 18cms. NR — BL —

Hanbüchlein der Lustfeuerwerkerei. Eine practische Anweisung zur Anfertingung aller Gegenstände der Lustfeuerwerkkunst, als: Schwärmer, Raketen, Räder, ... so wie vollständige Land= und Wasserfeuerwerke ... insbesondere aber für Dilettanten von Ruggieri.

R060.8
2e Aufl. G. Basse: Quedlinburg und Leipzig 1845
PC+TP+x+178+iiads.+29Pls. 20cms. NR — BL 8715.aaa.14

Handbuch der Lustfeuerwerkerei. Eine practische Anweisung ...

R060.9
3e Augl. G. Basse: Quedlinburg 1861
NR — BL —

Handbuch der Lustfeuerwerkerei. ... nach den neueren Werken von F.M. Chertier und L.E. Audot umgearbeitet.

R060.10
4e verb. Aufl. G. Basse: Quedlinburg 1864
xi+268+4Pls.(illus.). NR — BL —

The 1810 edition is not recorded anywhere except by bookseller Lucien Goldschmidt. The 'Vorwort' to the 2nd edition of 1845 is dated 1832 so obviously refers to the book entitled *Pracktisches Unterricht* .... The 29 plates are similar to those in Ruggieri's *Elémens de Pyrotechnie*.

References: Brock (1801[?], 1812, 1821). Du Pont. Kayser (All German eds.). CP (1811, 1812, 1821, 1830, 1845)

RUGGIERI, Claude Fortune and MOREL, Thomas

Die Pyrotechnie, oder theoretisch-praktische Anweisung zur Bereitung aller Arten Feuerwerke ... nach der Vorschriften von Claude Ruggieri und Thomas Morel.

R070.1
P.G. Kummer: Leipzig 1807
TP+xxiv+272+22Pls. 20cms. NP 0661841 BL —

References: Kayser. Brock

RUGGIERI, E.-F.-D

Catalogue des livres rares et precieux composant la bibliothèque de E.F.D. Ruggieri, sacre de rois et des empereurs, entrées triomphales, marriages, tournois joutes, carrousels, fêtes populaires et feux d'artifice.

R080.1
Labitte: Paris 1873
208p. 25cms. NR 0505084 BL 11905.i.35

Catalogues des livres anciens et modernes rares et curieux, fêtes publiques, entrées solomnelles, sacres, tournois, carrousels, ceremonies, feux d'artifice, pompes funebres, artillerie, pyrotechnie, etc. Composant de feu M. D.-E.-F. Ruggieri.

R080.2
A. Chossonery: Paris 1885

NR 0505082 BL S.C.1388

The above two auctioneers' catalogues are included in this bibliography because they form an important reference work which details the collection of the foremost European pyrotechnist of his day.

RUMYANTSEV, P

Teoreticheskaya i prakticheskaya pirotekhniya. [*Theoretical and practical pyrotechny.*]

R090.1
Moscow 1852

NR — BL —

References: SHIDLOVSKY's bibliography

S, J See SMITH, Capt. John

SAKENOBU, Noboru

Hanabi zakki. [*Miscellaneous notes on Fireworks.*]

S010.1
1970

NS — BL —

References: HOSOYA bibliography

SANDMANN, Max

Die Lustfeuerwerkerei zur Verschonerung offentlesser und hausliche Feste. Eine Anleitung zur Anfertigung samtlicher Ferwerkskorper nach dem heutigen Stande de pyrotechnischen Wissenschaft für Dilettanten wie Feuerwerker ... [*The making of fireworks for the embellishment of public and private celebrations. Instructions for making all firework items according to the present developements in the science of pyrotechnics for amateurs and fireworkers.*]

S020.1
10e. aufl. Bernard Friedrich Voigt: Weimar 1884
xii+234(64illus.). 23cms. NS — BL —

S020.2
11e. aufl. Bernard Friedrich Voigt: Weimar 1889
xii+234(64illus.). 23cms. NS — BL —

The preceding nine editions, all published by B.F.Voigt, are listed under BUETTNER who wrote the first in 1821 under the title *Die Feuerwerkerkunst für Dilettanten ... aus dem Franzosichen.* The French work in question being AUDOT's *L'Art de Faire a peu Frais les Feux d'Artifice.* BUETTNER'S work then went through numerous revisions and editorships incorporating excerpts from both CHERTIER and RUGGIERI, the 10th and 11th editions being normally listed under SANDMANN in the German Bibliographies.

References: Du Pont. Kayser. G&O

SAXON, Kurt

Fireworks and Explosives like Grandad used to make.

S030.1
Atlan Formularies: Eureka, California (1975)
61p+illus. 28cms. NS — BL —

Poor Man's James Bond. (K. Saxon ed.)

S030.2
Atlan Formularies: California 1977
146p. NS — BL —

References: NUC SC 1975

SCHARFENBURG, August

Die Feuerwerkskunst in ihrem ganzen Umfange. Lehrbuch der Lustfeuerwerkerei für Kunstler vom Fach und Dillettanten ... bearbeitet von A. Scharfenburg. [*The entire art of fireworks. Manual for making fireworks for professionals and amateurs. Edited by A. Scharfenburg. In two parts.*]

S040.1
E. Ebner: Ulm 1848/52
(I)TP+viii+448 (II)TP+345illus. 22cms. NS 0814737 BL —

S040.2
E. Ebner: Ulm 1854
NS 0814738 BL —

S040.3
(3 Ausgabe) Druck und J. Ebner'schen: Ulm 1865
682p. 22cms. NS — BL 8716.c.15

References: Du Pont (1865). Kayser

SCHMIDLAP, Johannes (von Schorndorff)

Künstliche vnd rechtschaffene Fewrwerck zum Schimpff, vormals im Truck nie auszgangen. Dermassen an tag geben vnd beschrieben deszgleichen mit artlichen Figuren dargethan das ein jeder so vormals solcher kein erfarung gehabt hat leychtlich sie ganz künstlich hierausz lernen mag. [*Artificial and honest fireworks for joy, never printed before. Published and described with the addition of artistic figures in such a way as to educate anyone even if he had no previous experience.*]

S050.1
(Johann von Berg vnd Vlrich Newber): Nürnberg M.D.LX. (1560)
TP(BR)+68L(32WC(12Full Page)) 16cms. NS — BL —

S050.2
1561
NS — BL —

H17 |£5000 **S050.3**
Johann von Berg vnd Vlrich Newber: Nürnberg 1564
vii+120L.illus. 15cms. NS 0231940 BL —

S050.4
Katherinam Gerlachin: Nürnberg 1590
TP(BR)+76L+36WC. 16cms. NS — BL —

S050.5
Katherinam Gerlachin: Nürnberg M D XCI(1591)
76L.illus. 16cms. NS — BL —

S050.6
Franckfurt 1602
NS — BL —

S050.7
P. Kauffmanns: Nürnberg MDCVIII (1608)
TP+76L+36WC 16cms. NS — BL 1438.b.13/2

Een ghetrou onderwiis van menigerhande Vyer-wercken. Om te maecken constige en aeridge Vyer-wercken van triumphe ende ghenoechte. [*A reliable instruction for many fireworks. To make artful and pleasing fireworks for triumph and amusement.*]

S050.8
Amsterdam 1624
NS — BL —

S050.9
Amsterdam 1630
NS — BL —

Lanthrop Harper in his catalogue described this as 'the earliest published work devoted exclusively to fireworks as a form of entertainment. Earlier books have dealt with pyrotechnics, ballistics, incendiary devices and artillery, but only from a military point of view. Schmidlap's text was probably first printed in 1561 and there were several reprints and a Dutch translation in the 17th-century; it is dedicated to Wilhelm von Januwitz, Master of Ordnance to the Duke of Württemberg. This appears to be the author's only published work, [COCKLE lists *De tormentis bellicis (1625)* under No.689], and few bibliographical details are known about him. The book is a complete guide to pyrotechnic displays, describing the manufacture of powder composed of saltpetre and sulphur, how to set up different kinds of firework and get the most amusing effects (smoke-candles, fire-crackers, explosive torches, the rotating wheel, a fire fountain, etc.); also showing how to build various rockets, and naturally how and when to take necessary precautions. At the end is a chapter on fireworks for masquerade during the annual carnival. Some illustrations are just instructive and diagrammatic, the larger cuts, though, are quaintly pictorial and rather interesting artistically; one of them represents two warriors wielding fireworks in the shape of weapons. The printer of the fascinating and pretty little book was the widow of Dietrich Gerlach and an active publisher in her own right of musical imprints.'

Sotheran also has a note on this important work in his *Biblotheca Chemico-Mathematica*. 'This would appear to be the first German work on Fireworks. In his preface the author mentions that the work is wholly original and states that he has refrained from treating more fully of fireworks for setting beleagured towns on fire which he very appropriately calls "Schimpfflick" (infamous) fireworks. He announces that he proposes to publish in a future work an account of the "flying dragon" which, however, does not appear to have come out.' Perhaps it appears in the Amsterdam edition which might not have been known to Sotheran. Lotz notes the existance of two Amsterdam editions but Cockle, quoting the *Cat. Lib.*

Manusciptorum Oxoniae 1697, lists only that of 1624 under the title *Pyrotechnia*.

The fact that the two forewords are both dated 1560 suggested both to Lotz and Harper that an earlier edition than was known to either might exist; it was therefore with great pleasure and excitement that I discovered this original 1560 edition in the Royal Danish Military Library, Copenhagen. Its title-page provides the frontispiece of this bibliography.

References: Cockle 934 (1608), 936 (1624). Du Pont (1591). Fahler (1591). Lotz (1591, 1602, 1608, 1624, 1630). Jähns (1561). RDML (1560, 1591)

SCHMIDT, Ivas

Nogle traek af Ildens og Fyraerkeriets Historie. Udgivet i Anledning af Ildfesterne paa Ingenierarfielleden i København. [*A brief history of fire and fireworks. Compiled in connection with the firework displays on the miliary parade grounds in Copenhagen.*]

S060.1
Copenhagen 1910
123p+illus. NS — BL —

References: RDL

SCHOTT, Gasper (1608-1666)

Magiæ Universalis naturæ et artis ... opvs qvaripartitum. Contines Pars. I. Optica. II Acoustica. III Mathematica. IV Pysica. [*The wonders of nature and art. 4 vols.*]

S070.1
J.G. Schönwetteri/H.Pigrin: Francofurti/Herbipoli M.DC.LVII. (1657)
(Vol.4)EngTP+xxiv+670+xviii+Pls. 20cms. NS 0272919 BL 535.e.8

Schott, like van ETTEN, was a prolific Jesuit author. For many years he was professor of Theology and Mathematics at Palermo as well as being an acknowledged expert on pyrotechnics. He wrote many works on Mathematics and Natural Philosophy, as well as compendium works on Science as ETTEN had done before him. All these books appeared in many editions; the NUC has 12 entries for *Magiæ universalis naturæ et artis* alone. BACKER comments that; 'la plupart des ouvrage sont curieux et ils étaient jadis fort recherchés; on y trouve des observations précieuses, des expériences dignes d'attention ...' The fourth volume of this four volume edition contains sections on cryptography, magnetism, medicine and a number of other subjects, as well as 117 pages on pyrotechnics. Acknowledgement is made to authors such as SIMIENOWITZ, UFANO, APPIER, FRONSPERGER and others.

SCHREIBER, Georg

Buchsenmeister-Discours. [*Gunner's discourse.*]

S080.1
Christoph Tschorn: Brieg (1656)
EngTP+TP(BR)+ii+Eng+xxiv+117+iii+77Pls+FT. 31cms.
NS — BL 718.k.34

S080.2
Christoph Tschorn: Brieg 1662. (EngTP 1656)
EngTP+TP+ii+Eng+xxiv+117+iii+77Pls+FT. 31cms.
NS — BL M.L.l.14

S080.3
Christoph Tschorn: Breslau 1671
117+78Pls. 32cms. NS 1277856 BL —

HODGKIN has this to say; 'Master Gunner discourse is an important work of 120 pages, profusely illustrated by etchings of Fireworks of both kinds. A conversation between a Zeugmeister (Artillery Officer), who wants to know all about everything, and the Buchsenmeister (Master Gunner), who is only too anxious to unburden himself of his stores of knowledge, runs through the book, and his explanations are made clear by the plentifully-interspersed illustrations, the Revolving Wheel of Fire gives a very good idea of a Catherine of the period, as shown in the larger prints of pyrotechnic displays, and the Eagle is as noble and as fierce a fowl as will be found among them all. Schreiber's bibliographic list of works on artillery down to the date of his own treatise comprises those of eighteen authors, including some familiar to us.' They include Christoff Manns (1578), Brechtel & Schmiedelop (Schmidlap) (1591), C.Clavii, J.Jacobi, de Bry, Dambach (1603), Fortenbach (sic) (1627), Adami Romani, J.Theodori, Uffani, J.Ammons, Ruscelli (1611). The editions are not identical as they have been re-set. There is a separate title-page for the section entitled *Feuerwerks Laboratorium* but paging is continuous. Plate numbers should be treated with caution as some copies have more than one figure on each plate, others have each figure on a separate plate.

References: Hodgkin

SCHROKA, D. G

Kurze Anleitung zur Feuerwerkerkunst ... bekannt gemacht von D. G. Schroka ... [*Brief instructions for fireworkers.*]

S090.1
J.F. Korn: Breslau und Hirschberg 1791
TP+vii+70+Pl. 20cms. NS — BL —

Kort og tydelig Beileduning til Fyrvaerkerie kunsten ... [*Brief instructions for the art of fireworks.*]

S090.2
Liunges: Copenhagen 1795
TP+x+88 16cms. NS — BL —

References: Berlin 3307 (1791). Keyser (1791). Barfod (1795)

SCHWENTER, Daniel (1585-1636)

Deliciae physico-mathematicae, oder Mathemat: und philosophische Erquickstunden.

S100.1
J. Dumler: Nurnberg 1636/53
NS 0329102 BL 716.f.3

Several editions were published between 1636 and 1677 by either J. Dumler or W.M. Endter in either 3 volumes or 3 in 1. The three parts consist of a translation by Schwenter based on ETTEN's *Récréations Mathematiques,* with two additional parts by Georg Philip Harsdorffern. These are Part II. *Deliciae mathematiciae et physicae* and Part III. *Deliciae philosophiciae et mathematicae.* Book XI of Part II has a chapter entitled *Von der Feuerkunst (Pyrabolia)* and occupies pages 428-463. Book VIII of Part III is entitled *Von der Feuer- und Schmaltzkunst* and occupies pages 498-524. The few woodcuts are not the same as those in the original work.

SHIDLOVSKY, A. A

Osnovy Pirotechniki. [*Fundamentals of Pyrotechnics.*]

S110.1
(1st ed.) Moscow 1943
248p. 22cms. NS 0504963 BL —

S110.2
(2nd ed.) Moscow 1954
284p. 23cms. NS 0504964 BL 8839.t.6

S110.3
(3rd ed.) Moscow 1964
338p. 23cms. RLS 095058 BL —

S110.4
(4th ed.) Mashinostroine: Moscow 1973
320p. NS — BL —

Foundations of Pyrotechnics. [*Trans. of 2nd ed.*]

S110.5
Wright Patterson Air Force Base: Ohio 1964
NS — BL —

Fundamentals of Pyrotechnics. [*Trans. of 3rd ed.*]

S110.6
U.S. Joint Pub. Research Services: Picatinny Arsenal 1965
NS — BL —

Basically a textbook of combustion science for university students similar to, but on a much larger scale than, DAVIS's was in America during World War Two. The American translation includes the unique and extensive bibliography of Russian firework books which I have included in this work.

SHIMIZU, Takeo (1912-)

Hanabi. [*Fireworks*]

S120.1
Hitotsubashi-Shobo: Tokyo 1957
341p. 19cms. NS — BL —

Hanabi no Hanashi. [*The History of Fireworks*]

S120.2
New Kawade Press: Tokyo 1976
220p. 19cms. NS — BL —

Feuerwerk vom Physikalischem Standpunkt aus. I Ausflage.

S120.3
Hower Verlag: Hamburg 1976
½T+TP+ii+252+ii+errata. 21cms. NS — BL —

Fireworks from a Physical Standpoint. Part I. [*Translation of the German edition of 1976.*]

S120.4
Pyrotechnica Publications: Texas USA 1981
ColPC+TP+Photo+vi+(1-67)illus. 28cms NS — BL —

Fireworks from a Physical Standpoint. Part II. [*Translation of the German edition of 1976.*]

S120.5
Pyrotechnica Publication: Texas USA 1983
ColPC+TP+(68-141)illus. 28cms NS — BL —

Fireworks. The Art, Science and Technique.

S120.6
Author by Maruzen Co.: Tokyo Japan 1981
½T+TP+Pl+xii+330incl.161figs. 27cms. NS — BL —

The author is without doubt the leading contemporary pyrotechnist. His detailed, comprehensive and wide-ranging analysis of all aspects of pyrotechnics places him head and shoulders above other technicians of his generation. His first book, published in in 1957, was the first systematic firework book printed in Japan. In 1972 a large and important part of it, the section dealing with the formulae and compositions of Japanese shells, was translated and published as part of LANCASTER's *Fireworks, Principles and Practice*, the first time such information became generally available to Western readers. *The History of Fireworks* (1976) was written more as a popular science type of book than a technical one. The various types of Japanese fireworks are described and also details of the various firework festivals that have taken place in Japan over the centuries, including the famous Ryogoku festival first held in 1733 as a thanksgiving and mass for those who died in the cholera epidemic of 1732. *Feuerwerk vom Physikalischem Standpunkt aus.* also published in 1976 was reviewed in *Pyrotechnica* No.IV by Jim Stone who described it as, ' ... a cross between Mount Everest and an encyclopaedia to the avid professional and amateur pyrotechnists', An English translation is currently being prepared, in three parts, by Pyrotechnica Publications. Two have so far appeared (Summer 1984).

References: CP (1976, 1976, 1981)

SI(E)MIENOWICZ, Casimir

Artis Magnæ Artilleriæ pars prima.

S130.1
Iaonnem Ianssonivm: Amsterodami MDCL (1650)
½TP+EngTP+xii+284+iv+22Pls. 33cms. NS 0537506 BL 64.f.1

Grand Art d'Artillerie. Mise en François par Piere Noizet.

S130.2
Iaonnem Ianssonius: Amsterdam 1651
EngTP+vi+41O+22Pls. NS 0537509 BL —

Vollkommene Geschutz = Feuerwerck = und Büchsenmeistery-Kunst: ... In die Hochtentsche Spraach ubersetzet: von Thoma Leonhard Beeren, Lipsiensi. Mit schönen Kupffern, und einem gantzen neuen Theil vermehret durch Daniel Elrich, Stückhaubtmann zu Franckfurt am Mayn.

S130.3
Henrich Friesen for Johann David Zunners: Frankfurt am Mayn 1676
EngTP+TP(BR)+x+232+iv+22Pls+EngTP+116+iv+25Pls. 31cms.
NS 0537512 BL 717.1.9

S130.4
(Facsimile reprint). Akademische Druck und Verlagsanstalt: Gratz 1976
½T+TP+viii+EngTP+TP+xii+232+iv+22Pls+EngTP+116+iv+25Pls. 28cms.
NS — BL X622/7206

The Great Art of Artillery of Casimir Simienowicz, Formerly Lieutenant General of the Ordnance to the King of Poland. Translated from the French by George Shelvocke, Jun. Gent.

S130.5
J. Tonson: London MDCCXXIX (1729)
EngTP+TP(BR)+xx+404+22FPls. 30cms. NS 0537501 BL 8825.f.27

S130.6
(Reprint) S.R. Publishers: England 1971
½T+TP+ii+EngTP+TP(BR)+xx+404. 27cms. NS — BL X622/968

Wielkiej sztuki artylerii czesc pierwsza. (Artis Magnae artilleria pars prima).

S130.7
Pol & Lat: Warsaw 1963
769p. NS — BL 08841.df.33

Russian edition with English commentary by A. Ivaskevicus.

S130.8
Moscow 1971
66p+12 illus. 26cms. NS — BL X6ll/3029

The following notes on the above editions are taken from London bookseller Paul Breman's excellent catalogue of Military Books issued in 1981, in which four different language versions were offered for sale.

'The title of this grand work reflects not its actual contents, but its author's intention to write a two-part Compleat Art of Artillery. Only the first book was finished when Simienowicz died. 'Das Werk ist vortrefflich angeordnet, klar in der Darstellung, i. allg. auf der Hohe der Zeit und hat daher bei seinem Erscheinen grosses Gluck gemacht; noch im 18. Jhdt. wurde es zu den Hauptwerken uber Artillerie gerechnet und stand im Tagesgebrauch. Es enthalt Jedoch ... fast nur die Feuerwerkerei.' (Jahns 1196-8). Indeed the five parts of the treatise deal with calibre, pyrotechnics, rockets, fire-balls, and 'warlike firearms' (in Riling's curious phrase; firework set-pieces would be a closer approximation of what Simienowicz, a long-winded bore, calls 'machines de guerre fixes ou mobile, tant recreative que serieuses ou militaires'). Probably precisely because it remained confined to 'recipes' the book retained its status of ultimate authority for a very long time. In 1676 Daniel Elrich added to the German translation a continuation of his own; at the turn of the century Gruber used some of the plates for his Kriegs-Schule; in 1729 George Shelvocke published an English translation, by order of the Surveyor-General of the Ordnance; and in 1741 Frezier drew rather liberally on the work for his *Traité des Feux d'Artifice*. 'It then still remained the outstanding work in both the military and civil branches of the subject' (Brock, *History of Fireworks*). The long and detailed section on rockets, at Simienowicz's time a completely obsolete weapon (?), is very curious and interesting. His war rockets go up to about 100lbs, and the plates devoted to them are rivalled only by the oft-plagiarized one of set-piece construction.

There is just one slight difference between the original Latin edition and the almost simultaneous French version by the same publisher: the first has a long explanatory preface to the reader and only a short formal dedication (to William Frederick von Nassau), the second has a long address to the dedicatee and no other preface.'

Of the German version Breman continues: 'The first part is the only (and quite faithful) rendering of Simienowicz's original into German; the title and all other engravings were printed from the original coppers. (Berlin Cat.3297. Lotz 73). Elrich's continuation in no way resembles Simienowicz's announced intentions for a second part. Instead, it elaborates on part one in a somewhat archaic, often also quite fascinating way. Elrich's description of a powder mill is accompanied by two plates, of which one affords views of both the inside and the outside by means of a dramatic over-slip. The preceding plate, of saltpetre manufacture, is equally interesting. Book five contains detailed instructions for building the grandest firework set-pieces, a burning castle: the principles are derived from Simienowicz's plate W but the design is infinitely more ambitious. In Books six and seven Elrich gets involved at last with one aspect of Artillery, the bombadier's art, with fine illustrations of mortar and petard.

The English version is in its strangely elusive and only edition. The 1600-item Maggs catalogue *Four Centuries of Military Books* (Cat.915) boasted only four works more expensive ... The translator has done an unusual amount of home-work, apart from being quite obviously well versed in the actual subject matter. He gives the sources for all of Simienowicz's many classical references and quotations (he even completes the mutilated final quotation, from Scaliger) and adds a few notes based on his own experience (wooden trumpions, for instance, 'are found to be useless'). The Plates for this edition were engraved by Pierre Fourdrinier. They are very faithful and not reversed; where necessary the captions and explanations have been translated. On the title engraving the dress of many of the figures has been modernised and anglicized. An advantage over the original is that the plates have been arranged to fold outside the book for easier reference.'

The Russian edition of 1971 is a translation of Simienowicz's chapters on Rocket construction. The preface deals briefly with the previous history of Rockets and the English commentary outlines Casimir's achievements in Rocketry and claims him as 'the Lithuanian inventor of the rockets of the 17th century'. Some of the plates from *Artis Magnae Artilleriae* are reproduced. The 1971 reprint was reviewed in *American Pyrotechnist Fireworks News* , September 1973.

References: Du Pont (1650, 1676, 1729). Brock (1650). Berlin 3297 (1676). CP (1650, 1676, 1729, 1971)

SIMONENKO, P

Pirotekhniya

S140.1
Moscow 1894
NS — BL —

References: SHIDLOVSKY's bibliography

SINCERUS(I), Alexander

Der wohlerfahrene Salpetersieder und Feuerwerker oder bende aufgefühet in einem ganz neu herausgegebennen Tractatlein ... von einem erfahren Feuerwercker zusammen getragen und ... am Tag gegeben, durch Alex Sincerum. [*The experienced saltpeter screener and fireworker.*]

S150.1
Johanu Adolph: Franckfurt 1710
TP+xiv+190. 17cms. NS 0572602 BL 1607/6018

S150.2
(3 Aufl) C.G. Hauffe: Frankfurt und Leipzig 1773
162p. 19cms. NS — BL —

S150.3
Frankfurt 1778
NS — BL —

References: Du Pont (1773). Brock (1710). Kayser. SOT (1778)

SMITH, Godfrey [NUC-George]

The Laboratory; or, School of Arts. In which are faithfully exhibited, and fully explained ... Translated from the German.

S160.1
T. Cox: London 1738
iv+242+Pls. 20cms. NS 0632746 BL 1651/13

S160.2
(lst ed) John James: London MDCCXXXIX (1739)
EngP+TP+242+lPl. 20cms. NS 0632747 BL 1490.b.42

S160.3
2nd ed. to which is added an appendix. J. Hodges, J. James & T. Cooper: London MDCCXL (1740)
EngP+TP+vi+240+3Pls+TP+lxxxincl.3FPls+viii. 20cms.
NS 0632748 BL 1651/11

The Laboratory; or, School of Arts. In which are faithfully exhibited, and fully explained ... VII The Art of preparing Rockets, Crackers, Fire-Globes, Stars, Sparks &c for Recreative Fire Works.

S160.4
3rd ed. J. Hodges, T. Ashley: London 1750
EngP+TP+vi+352+11Pls(2F)+viii. NS 0632749 BL 1651/15

S160.5
4th ed. J. Hodges: London MDCCLV (1755)
EngTP+TP+xiv+352+9Pls+2FPls. 22cms. NS 0632751 BL 1509/919

The Laboratory; or, School of Arts: ... with ... a complete treatise on Fireworks ... [2 *vols.*]

S160.6
(Vol.II). C. Hitch ... : London 1756.
NS 0632753 BL C108.aa.15

S160.7
5th ed. S. Crowder ... : London 1770
EngTP+TP+352+Pls 22cms. NS — BL 7939.c.56

S160.8
6th ed. C. Whittingham for H. Symonds ... : London 1799
44O+15Pls+464+16Pls. NS 0632754 BL 7953.f.35

S160.9
5th ed. J. Seddon: Manchester [1800]
20cms. NS 0632755 BL —

S160.10
7th ed. Sherwood, Neely & Jones ... : London 1810
2 vols Pls. 20cms. NS 0632756 BL 1400.e.10

For some perverse reason the NUC lists this work under 'George' rather than 'Godfrey' despite the signature of 'Godfrey Smith' to the dedication. The first edition of 1738 does not contain any reference to Fireworks, although it does contain a section dealing with the manufacture of Saltpetre. The 'Second Edition' of 1740 however has an 'Appendix' with separate title-page, which includes, apart from a section on 'Dyeing', 'Proper lessons for preparing and managing all sorts of Rockets, Crackers, Fire Wheels, Fire Globes, Balls, Stars, Sparks etc. in the executing of artificial and recreative Fireworks. Translated from the High Dutch'. There was another issue of the 2nd edition of 1740 (BL1486.r.13 & C.108.aa.15) in which the first part and index were completely reset. The appendix, with separate titlepage, pagination and register is a reissue of the earlier 1740 'Second edition', with the title-page partially reset, removing the statement of price and substituting Roman for Arabic numerals in the date (ESTC). This section contains 80 pages in all and three additional plates. The subsequent editions all contained this firworks section and even 70 years later, the plates, containing nearly 70 figures, were, with only three small exceptions, identical.

References: CP (1755 & 1810)

SMITH, Capt. John (Governor of Virginia). (1580-1631)

The Seaman's Grammer.

S170.1
Andrew Kemb: London 1653
TP+85p. NS — BL E679/9

The Sea-man's Grammer and Dictionary ... with the com-positions for the making of several Fire-works useful in War both at Sea and Land.

S170.2
Randal Taylor: London MDCXCI (1691)
TP+iv+163illus.+1FPl. NS 0640796 BL 51.c.8

Only half a page of the 1653 edition is devoted to the subject of Fire Arrows, Pikes, Wild Fire and Granados. By the 1691 edition this had been expanded to 9 pages relating to Fireworks and their construction. There is no suggestion that they can be used for recreative purposes. Other editions and printings exist. For full details of all of Capt. John Smith's work, see his Bibliography by Wilberforce Eames, reprinted from Sabin's *Dictionary of Books relating to America*, New York 1927. See also note under the next entry.

References: Wing S4123-6

[SMITH, Capt. John] (Late of London, Merchant). (c.1633-1673)

Profit and Pleasure united, or the Husbandman's Magazene, being a most exact treatise of Horses ... Oxen ... and all other domestick cattle ... To which is added, the art of angling ... ringing, And making Fireworks. The whole elusterated with copper cuts. By J.S.

S180.1
Printed for S. Lee & Andr. Thorncome: London 1684
TP+ii+210+4Pls. 14cms. NP 0599692 BL D

Why do so many John Smiths have to use 'J.S.' as their pseudonyms and why do several of them have to be 'Captain' and live at about the same time? This problem is not helped by the way the NUC and BL catalogue list them. Captain John SMITH, Governor of Virginia, was the author of 'The Seaman's Grammer'ı and is listed under the previous 'SMITH' entry. He is sometimes listed as plain 'John SMITH' by NUC, and the fact that several editions of his 'Seaman's Grammers'ı were issued after his death and therefore contemporaneously with 'Profit & Pleasure'ı can be confusing. However the author of the latter work was a certain Captain John SMITH, a 'Merchant of London' as we are informed by the BL catalogue, whose main claim to fame was a book called 'England's improvement reviv'd'ı which, like 'Profit and Pleaure'ı seems to be a small encyclopaedia! The only UK location is the Royal Veterinary College library. The 'Tretis of Artificial Fireworks for Recreation', (spelling was not one of Smith's strong points), occupies pages 190-201 between a chapter on Bell Ringing and a list of diseases of Oxen. These pages deal with 'Skye-rockets', 'Ground Rockets', 'Golden Raine', 'Saussicons', 'Girondels' and 'Ballouns'. One can but hope that his veterinary advice is more practical than his ill-spelt, ill-written 'tretis'.

I am indebted to WING for helping to resolve much of the above although it still leaves the problem as to which 'J.S.' is responsible for 'The SOULDIERS Companion'ı (1688); is he the same as the Capt J. S. (yes, another one!) the author of *Military Disipline*.

An American edition of *Profit and Pleasure* is listed by NUC under NS 0640631 published in 'Boston in New England' in 1718. The long-title as quoted however does not include the phrase, 'And making Fireworks'; had it done so it might lay claim to the first American book to deal with the subject.

References: Wing S4095. Arber

SMITH, Thomas (c.1600-1627)

Certaine additions to the booke of Gunnery, with a supply of Fire-workes.

S190.1
William Ponsonby: London 1601
NS — BL —

S190.2
H.L. for R. Dawlman: London 1627
76+1FPl+illus. NS 0652764 BL 1399.c.2

(Eds 1628, 1643.)

Contains 56 pages on military fireworks, mainly Lances, Arrows, Fireballs etc for 'Sea and Land service'. Despite the long-title claiming the 'additions' to be 'both pleasant and profitable', in the last paragraph the author states, 'I might here adjoin variety of serviceable Fireworks, as also show many devices of pleasant Fireworks to be used in triumph before great personages; but to the wise these may suffice, who by experimenting these may easily perform the other.'

In the BL copy of 1627, the title-page which encloses both *The Art of Gunnery* and *Certain Additions* reads, *The Complete Souldier. Containing the Whole Art of Gunnery, with certain new and rare additions concerning Fire-works.*

References: STC22855, 22855.5, 22856.

SOLODOVNIKOV, B. M

Piroteknika [*Production and Combustion of Fireworks.*]

S200.1
Oborongiz: Moscow 1938
NS — BL —

References: ELLERN

SONZOGNO, Cesare

Il Pirotecnico Moderno che insegna l'Arte di Fare i Fuochi d'Artifizio ... [*Modern pyrotechnics which teaches the art of making fireworks.*]

S210.1
Lorenzo Sonzogno/Corso Francesco: Milano 1846
PC+TP+150(-4)+ii(incl.13F.Pls). 16cms. NS 0729098 BL —

S210.2
Seconda ed. riv. et aum. Ronchetti: Milano 1858
147+13Pls. NS 0729099 BL —

S210.3
3rd ed
NS — BL —

S210.4
Quarta ed. riv.et.au. Paulo Carrera: Milano 1877
PC+½T+TP+128(-4)+13FPLs. 18cms. NS 0729101 BL —

S210.5
Quinta ed. riv. et aum. P. Carrera: Milano 1892
189p. 18.5cms. NS 0729102 BL 8716.aaa.43

La Pirotecnia dei dilattante. Ed. Sonzogno.

S210.6
Milano 1900
62p. 16mo. NS — BL —

References: Pagliaini (1858, 1892, 1900). CP (1846, 1877)

SOULDIERS

The Souldiers Companion: or, Military Glory display'd, in a true and impartial description of all the memorable battels and fights by land and sea ... Together with the art of Gunnery, and preparing Artificial Fire-works for war or recreation ... by J.S.

S220.1
Nath. Ponder: London 1688
TP+vi+252. 14cms. NS 0741732 BL —

Possibly by the same 'Capt. J. S.' who was the author of *Military Discipline, or the Art of War* 1685?/1689. Only the last ten pages deal with fireworks. Not in Halkett & Laing. See also note under John SMITH.

References: Wing S.88. (Wing lists it spelt as 'Soldiers'). Du Pont. Arber.

SPLIT, M See STEERK, Major

SPON, Ernest

Workshop Receipts, for the use of Manufacturers, Mechanics, and Scientific Amateurs.

S230.1
E. & F.N. Spon: London 1875
TP+ii+450 NS — BL 7949.l.17

24 pages are devoted to several hazardous Firework recipes containing chlorates as well as recipes for Guncotton and Nitroglycerine. This section follows that on Plástering and Whitewashing! Several editions and re-issues appeared from 1873 onwards.

STARKEY, George (1627-1666)

He wrote a number of books with 'Pyrotechny' in their title such as *La Pyrotechny Starkey* and *Pyrotechny Assorted* but these deal with matters such as Alcohol and Pharmacy *not* fireworks!.

STEERK, Major

Guide practique de la fabrication des poudres et salpetres ... suivi d'un appendice sur les feux d'Artifice par M. Split.

S240.1
Lacroix: Paris c.1866
½T+TP+356(incl.12figs). 18cms. NS — BL 7943.aaa.37

Pages 321-356 contain M. Split's 'appendice'.

STEPANOV, P. V

Pirotekhniya, kurs feyerverochnogo iskusstva. [*Pyrotechny, a course in the art of fireworks.*]

S250.1
St. Petersburg 1894
NS — BL —

References: SHIDLOVSKY's bibliography

STOEVESANDT, J. C

Deutliche Anweisung zur Feuerwerkerey, worinnen alle gebräuchliche Arten der Lust- und Ernstfeuer nebst derselben Berfertigung und denen dazu nötigen Werkzeugen ordentlich und genau beschrieben und mit Kupfern erläutert werden. [*Simple instructions on making fireworks, in which all the usual types of recreative and military fireworks together with the manufacture of the same and the tools needed are precisely and accurately described and illustrated with copper plates.*]

S260.1
J.J. Gebauern: Halle 1748
95+13Pls. NS — BL 717.h.11

S260.2
Q922 Johann Friedrich Gleditschens: Leipzig 1757
TP+x+84+13FPls. 23cms. NS 0960395 BL 1651/1254

S260.3
Neue Auflage. Gleditschens: Leipzig 1778
FPl+TP+x+96+12FPLs. 23cms. NS — BL —

1784? 15R88

Sotheran says: 'Gives a full account of how to make fireworks, with a description of all the necessary tools.' HODGKIN also considers it to be a 'useful and practical little work containing the most explicit instructions for the manufacture of a great variety of pieces. The composition of a variety of powders for various effects is given in great detail, and, all the dimensions of every portion of the various pieces being furnished, nothing is left to the invention or caprice of the amateur or manufacturer. Firework making has become an exact science. The illustrations give as lively a presentiment of rocket-making in the middle of the eighteenth century as that of Van Etten in the beginning of the seventeenth ...'

References: Berlin 3303. Kayser gives 1748 & '56 eds. CP (1757, 1778)

STRUENSEE, Carl August

Anfangsgründe der Artillerie ... [*Fundamental principles of artillery.*]

S270.1
David Siegertischen: Leipzig 1760
TP+xxx+496+xvi+29Pls. 20cms NS — BL —

Contains a few pages and plates dealing with fireworks

References: Barfod

STUBBE, (or STUBBS), Henry (1632-1676)

Legend no Histories ... wherein ... sundry mistakes about the making of Salt-petre and Gunpowder are detected and rectified ...

S280.1
London 1670
127p. 22cms. NS 1020346 BL 740.c.19

STURMY, Samuel (1633-1669)

The Mariners Magazine, stor'd with these Mathematical Arts: ... Gunnery and Artificial Fire-Works.

S290.1
E. Coates: London 1669
EngTP+TP(BR)+528+11Pls(10F). 30cms. NS 1030575 BL L32/39

S290.2
2nd ed. Anne Godbid for William Fisher ... : London 1679
471+12Pls(8F). 30cms. NS 1030579 BL —

S290.3
3rd ed. rev. & cor. (Revised by John Colson) John Playford for W. Fisher etc.: London 1684.
Pl+EngTP+TP(BR)+82+TP+85-196+TP+106+109-174+TP+20+TP+22+Pls.
30cms NS 1030580 BL C.31.m.18

S290.4
4th ed. rev. cor. & enlarged by John Colson. Richard Mount: London 1700
2 vols. in 1. 30cms. NS 1030582 BL C.115.g.1

The first engraved title-page has the general title *Sturmys Mathematicall and Pracaticall Arts*. The collation is a very confusing collection of 'Books' and

'Chapters' some with and some without their own title-pages. 'Artificial Fireworks' occupies Book V Chapter XIII pages 83-90 and one plate. The author acknowledges that his material has been derived from Babington, Bate, Norton and Malthus. It contains all the things that a Master Mariner was expected to know and includes chapters on Navigation, Geometry, Instruments, Surveying, Astronomy, Dialling and much else.

References: Wing W6096-8. CP (1684)

STUWER, Anton

Der kleine vollkommene Lustfeuerwerker. [*The complete small fireworker.*]

S300.1
J. Benedikts Witwe & Sohn: Wien 1847
NS — BL —

References: Pokorny's collection

SYLVIUS, Frederick (Duke of Würtemberg und Teck)

Unterschiedene neue Arten von Künstlichem Fewerwerck neben Kurtzembegrief vnd anleitung zu der Artillerie. Von einem Liebhaber beyderley Wissenschafft hiebevor aufgesezet iezo aber Anderen zufernerem nachsinnen an tag gegeben. [*Different new types of artificial fireworks together with a brief concept and instruction on artillery.*]

S310.1
(Johann Seyffert: Oelsse) M.DC.LVII. (1657)
EngTP+iv+72+iv+51Pls. 29cms. NS 1113642 BL 8715.k.7

Vollenkommene Vnterweisung, wie Raketen, Feuer- Wasser- Sturm-Kugeln, Granaten, Pech-Sturm-Krantze, und allerhand lust nnd Ernsthaffte Feuer-werke zubereiten. Sampt gründlicher Anleitung zur Artillerie: als vom Mortier, von allerhand groben und kleinen Geschötzen, wie solche zu probiren, zu laden, zu richten u.d.g. Benebenst vielen dazu nötig dienlichen raren und fast unbekanden Instrumenten, von einem Liebhaber beyderley Wissenschafften an Tag gegeben ... [*Complete instructions on how to make rockets, fire, water and storm balls, grenades, pitch and storm wheels and all kinds of recreative fireworks.*]

S310.2
Tilman Bucholtz: Ossnabrück 1660
½T+TP+iv+71+5lPls. 31cms. NV — BL —

The BL listing is under SYLVIUS NIMROD, but NUC quotes Jähns who states that the book was probably written by Frederick 'Hertzog zu Würtemberg und Teck; auch in schleisen zur Oels' although the 'Vorrede' is signed 'SYLVIUS Hertzoz zu Würtemberg und Teck; auch in schlesisen zur Oels.' The author is not acknowledged in the 1660 edition as it was reprinted at the request of a nobleman as the original was out of print. The half-title is called 'Praxis Artolloriae Pyrotechnicae'. The book is divided into two parts, the first headed, 'Von bereitung des Lust- und Ernst-Fewr' and the second, 'Ander der Artollery'. A large section of the first part and some half-dozen plates deal with wheels rockets and shells. In BLONDEL's bibliography he quotes an edition of 1663 as well.

References: Du Pont (1660). Berlin 3298. RDL (1660). CP (1657). Hodgkin (1657)

T, W See VENN, Thomas

TARTAGLIA, Nicholas See LUCAR, Cyprian

TESSIER, Paul (1816- ?)

Chimie Pyrotechnique ou Traité Pratique des Feux Colorés.
[*Pyrotechnic chemistry or practical treatise of coloured fire.*]

T010.1
J. Dumaine, Lacroix et Baudry: Paris 1859
½T+TP+x+438. 22cms. NT 0111616 BL 8715.c.38

Chimie Pyrotechnique, ou Traité Pratique des Feux Colorés. Deuxième édition entirèment réfondue, et augmentée de quelques Nouveaux Artifices.

T010.2
L. Baudoin et Cie: Paris 1883
½T+TP+xviii+458. 22cms. NT 0111617 BL —

One of the foremost theoretical pyrotechnic chemists of his time. Many of his formulae are basically still in use today. He was not, however, the intuitive practical pyrotechnist that CHERTIER was.

References: Brock (1859). Du Pont. CP (1859, 1883)

THYBOUREL, Francois See APPIER, Jean

TÖNER, A.R

Fyrverkerikonsten med hufvudsakligt afseede på Lustfyrverkeriet. [*The art of firworks with the principles of recreative fireworks.*]

T020.1
S. Flodins: Stockholm (1885)
TP+xiv+388inc.129figs. 19cms. NT — BL —

References: Barfod

TRATTATO TEORICO-PRACTICO

Trattato teorico-practico Pirotechnico ossia l'Arte di fare i Fuochi di Gioia. Contenete Il dettaglio generale e particolorizatio della materie, degle utensili, e della maniera di construili, dollocarli e adpoerarli. [*A theoretical and practical treatise on pyrotechnic or the art of making fireworks for celebration.*]

T030.1
Faziola: Genova 1853
250p+28FPls. 4to. NT 0312591 BL —

A fairly complete translation of RUGGIERI's Elémens de Pyrotechnie(1821) complete with reversed plates.

References: MS.

TRINKHAUS, E

Kunst und Lustfeuerwerkerei.

T040.1
Hennings und Hopf: Erfurt 1841
PC+TP+xvi+136+2FPls. 13cms. NT — BL —

References: Kayser. RDML

TRUE

The True and Easy Guide to Conjuring ... To which is added the Art of Making Fireworks.

T050.1
Stevens & Co. ... : London c.1815
vii+8-28p. 18cms. NT — BL —

T050.2
New edition. G. Stevens ... : London c.1816
Col.FP+32. NT — BL —

References: Stott HP266. Hall 275/6

TSYTOVICH, P. S

Opyt ratsional'noy pirotekhnii. [*Experiment in practical pyrotechny.*]

T060.1
St. Petersburg 1894
NT — BL —

References: SHIDLOVSKY's bibliography

UCHATIUS, Josef von

Die Kunst-Feuerwerkerei zu Lande. Basirt auf wissen-schaftliche Principien, und mit besonderer Berücksichtigung einer ... Anweisung zur Verfertigung sowohl einzelner Feuerwerks-stucke als ganzer Feuerwerke jeder Grösse ... [*The making of land fireworks. Based on scientific principles. ... Instructions for making both individual firework items and complete fireworks of every size.*]

U010.1
F. Tendler: Wien 1848
TP+xxx+482+ii+14FPls+FTabs. 22cms. NU — BL 1399.h.2

Superbly detailed plates and descriptions of set-pieces which also includes the earliest account of electrical firing of fireworks.

References: Du Pont

UFANO, Diego

Tratado de la artilleria y vso della praticado en las guerras de Flandes. [*Treatise on artillery and the use made of it in the Flanders war.*]

U020.1
Ivan Momarte: Brvsselas 1612
EngTP+xvi+424+viii+55Pls. 31cms. NU — BL 1480.bb.33

U020.2
Ivan Momarte: Brvsselas 1613
EngTP+xvi+424+viii+ Pls. NU 0005447 BL —

U020.3
Ivan Momarte: Brvsselas 1617
EngTP+xvi+424+vii+ Pls. NU — BL —

Artillerie c'est a dire ou vraye instrvction de l'Artillerie et de tovtes ses Appartenances. ... avec vn enseingment de preparer toutes sortes des feux artificels, tant pour resouyer les amis, qui pour molester & endommager, & par eau & par terre les ennemis.

U020.4
Engolf Emmel: Franckfort MDC.XIV (1614)
EngTP+Eng+v+164+iv+26Pls. 31cms. NU — BL C47.i.12/1

U020.5
Andre d'Aelst: Zvtphen 1621
TP+ii+144+6+26Pls. 31cms. NU 0005446 BL ML.f.10

U020.6
Jean Berthelin: Rouen 1628
iv+144+6+Pls. 31cms. NU 0005448 BL —

U020.7
Jean Iansson: Amsterdam M.DC.XXXIV (1635)
EngTP+144+6+Pls. 31cms. NU — BL —

Archeley, das ist Gründlicher vnnd Eygentlicher Bericht von geschütz vnd aller Zugehör.

U020.8
E. Emmeln: Franckfurt MDC.XIV (1614)
EngTP+Eng+ix+180+27Pls. 31cms. NU — BL 1602/123

U020.9
Erasmo Kempffern: Franckfurt MDC.XXI (1621)
EngTP+Eng+iv+132(-7)+ii+26Pls. 31cms. NU — BL 1605/217

U020.10
Andries: Zvtphen M.DC.XXX (1630)
EngTP+ii+128+ii+6+29Pls. 31cms. NU — BL ML.p.8

Archelia albo Artilleria to iest Fvundamentalna y Doskonala informacya o Strzelbie, y o rzeczach do niey nalezacych ...

U020.11
J. Dekau: Lesznie M.DC.XLIII (1643)
EngTP+TP+xii+160+72+iv+60Pls. 31cms. NU — BL 8824.h.33

UFANO's *Tratado* provided source material for a very great number of subsequent works and his original plates, re-engraved (and reversed in the process) by Theodor de BRY, are used in all the other editions. De BRY also made the translations for the French and German editions. There is an edition of 1612, as well as of 1613, although COCKLE does not mention the 1612 edition and the entry in Palau y Dulcet says, 'Almirante (*Bibliographia militar de Espana*, Madrid 1876.) believes that the edition of 1612 cited by Rios y Salas does not exist, but Garcia Rico gives Brussels 1612 without details. Almirante also denies the Ambrese edition of 1613 cited by Huerte y Lacuce.' This ignorance of the 1612 edition seems surprising since the dedications are dated October 1611 and January 1612. Full details of all the above editions (except the Polish) are to be found in Palau y Dulcet. No two authorities agree as to the number of plates, so any figure must be treated with caution. (Palau y Dulcet gives the 1613, 1614, 1621 & 1628 editions 47, 28, 35 and 29 plates respectively). This confusion is caused by some of the plates, consisting of more than one figure, having been cut before collation in the appropriate place in the text. In other versions the plates have all been bound in at the end of the text.

Among those who have made use of UFANO's work are ELDRED in his *Gunner's Glasse* of 1646, although he makes no mention of Fireworks. NORTON's *Gunner* not only paraphrases the text but also uses the de BRY plates with the excuse that, 'by reason of their goodness, fairness and cheapness, for the Figures, had they been cut of purpose in England, would have made these bookes too deare for Gunners'. VENN's *Military and Maritine Discipline* Part III (1672) also makes use of UFANO.

Of the book's 30 chapters the last five are devoted to fireworks 'de joye' and include descriptions of horizontal and vertical wheels, spheres, flying dragons, line rockets and a 'machine' depicting a chateau.

References: Brock(1612). Cockle 684. Palau y Dulcet. RAI (1635)

UPMANN, Dr. J

Das Schiesspulver, diesen geschicte, fabrikation eigens-schafter und proben. [*Explosive powder, its history, manufacture, properties and tests*. (From Bolley's - *Handbuch der Chemischen Technologie, Bd.6. Gruppe 3. Abt.1.*)]

U030.1
F. Vieweg und sohn: Braunschweig 1874
viii+219illus. NU 0288538 BL 8905.eee.

For the French translation of this together with E. von MEYER's work, *Die Explosivkorper*, which forms the second section of 'Gruppe 3.', see DESORTIAUX *Traité sur la Poudre.*

References: Du Pont

VALLE, Battista della (? -1535)

Vallo Libro Co(n)tinente ...

V010.1
np. Venetia M.D.XXIIII. (1524)
EngTP+xiv+72incl.WC. NV 0020046 BL 1607/961

Eds.1528, 1529, 1531, 1535, 1543, 1545, 1550, 1558
French eds. 1529, 1531, 1554

Printed on the verso of the last page of the table of contents is an engraved plate depicting trombe, fire-pots and bomb. The ensuing 6 pages (not included in the table of contents) are headed, 'Capitoli tre de Artificii de fuoco nouamente aggionti'. The following three sections describe 'Trombe', 'Balle de Bronzo' and 'Fiaschi'. Hodgkin comments that this probably represents 'the first printed description of military fireworks of this kind'. His edition was 1529. There is however no mention of recreative fireworks.

References: Hodgkin. Lotz

VASILEV, Vladamir Nikolaevich

Starinnye Feierverki v Rossi: XVII Pervaya Chetverth XVIII Veka.

V020.1
Izdatetelstovo Gosudarstvennogo Ermitajhka: Leningrad 1960
55+ 31Pls. NV — BL 8841.d.10

Concerns firework displays given in Russia in the 17th and 18th centuries.

References: BC

VENN, Captain Thomas

Military and Maritine Discipline in three books. Book I. ... Book III. The Compleat Gunner ... shewing the art of founding of great ordnance; making of gun-powder ... with the nature of Fireworks. Translated out of Casimir [Simienowicz], Diego, Vffano and Hexam &c. (By W.T.). [3 Books in one.]

V030.1
Printed by E. Tyler and R. Holt for Rob. Pawlet (etc.): London 1672
EngTP+TP(BR)+TP+xiv+206+ii+(II)EngTP+TP+ii+68+9Pls+
(III)TP+ii+(88+26)+4Pls+7TP+15Pls. 30cms.
NM 0577570 BL 534.m.8

The Compleat Gunner, in Three Parts. Part I. ... The composition and Matters of Gunpowders, the several Sorts, Colours, and Operations. Part II. ... Part III. Shews the Nature of Fire-works, the manner of composing many that are excellent and useful both for Sea and Land ... Translated out of Casimir, Diego, Uffano, Hexam and other Authors. To which is added the Doctrine of Projects ... (By W.T.).

V030.2
Rob. Pawlet, Tho. Passinger, and Benj. Hurlock: London 1672
TP+ii+88+26+4Pls+WC. 30cms. NC 0600553 BL —

V030.3
(Reprint) S.R. Publishers: (Wakefield) 1971
(½T+TP+iv)+TP+ii+88+26+4Pls+WC. 23cms.
NC — BL X620/2157

Thomas VENN is credited only with the first of the three books that make up *Military and Maritine Discipline*. The third book is by 'W.T.' a pseudonym not mentioned by Halkett & Laing. This 'Book' was separately published and has recently been reprinted in a slightly smaller format. Despite its long-title, it does include 12 pages devoted to 'Artificial Fire-works for Recreation'. The reader is invited to study the works of Siemienowitz, Bate, and Babington for fuller details, but nevertheless he describes rockets and their composition, stars, rain, serpents, fire-wheels together with 22 rules to be complied with 'by which you may make fuzees or Rockets, without any default'. The four plates (in the Artillery section), are almost identical to those that NORTON (1628), and others, lifted from UFANO's *Tratado* of 1612, (not 1613 as Brigadier Hogg says in his introduction to the reprint).

References: Du Pont. WING V192. CP (Reprint). Arber

VERGNAUD, A(r)mand Denis (1791-1885)

Manuel de l'Artificier, ou l'Art de faire toutes sortes de Feux d'Artifice a peu de frais, et d'apres les meilleurs procedes; contenant les élémens de la pyrotechnie civile et militaire, leur application practique a tous les artifices connus jusqu'a ce pour, et a de nouvelles combinaisons fulminantes. [*Fireworkers manual, or the art of making all sorts of fireworks at little cost, and after the best practice.*]

V040.1
Roret: Paris 1826
xiii+317+FPl. 14cms. NV 0107393 BL —

Manuel de l'Artificier, du Poudrier et du Salpêtrier ...

V040.2
Paris 1831
NV 0107424 BL —

Nouveau Manuel de l'Artificier, du Poudrier et du Salpêtrier, contenant les élémens de la pyrotechnie civile et Militaire, ... Nouvelle edition, entièrement refondue et considérablement augmentée.

V040.3
Roret: Paris 1838
½T+TP+ii+344(-6)+2FPls+5FT. 15cms. NV 0107423 BL 12207.a.1/41

V040.4
(Reprint) Leonce Laget: Paris 1981
½T+TP+ii+344(-6)+2FPls+5FT. 17cms. NV — BL —

Nouveau Manuel Complet de l'Artificier, du Poudrier et du Salpêtrier, contenant ...

V040.5
Roret: Paris 1852
vi+430+Pls+tbs. NV 0107411 BL —

VERGNAUD, A.D. & VERGNAUD, P

Nouveau Manual complet de l'Artificier. 1 ere Partie: Pyrotechnie Militaire. 2 eme Partie: Pyrotechnie Civile.

V040.6
Roret: Paris 1865
(I) ½T+TP+ii+522+3FPls+5FT. (II) ½T+TP+iv+204+2FPls.
NV — BL 12207.a.1/336

V040.7
L. Mialo: Paris 1906
464p+Pls. 16cms. NV 0107408 BL 12207.a.1/544

Manual elemental de pirotecnia civil y militar ... 3rd ed. aumentala con la Pirotecnia popular ...

V040.8
Cuesta: Madrid 1856
xvi+231+FPl. 16cms. NV 0107398 BL —

The vast number of editions that were brought out under the imprint of 'ENCYCLOPEDIE-RORET' and 'MANUELS-RORET' make it very difficult to separate the various versions. This is particularly so when the NUC sometimes only gives abbreviated titles and some hesitant dates. (The NUC quotes an 1828 edition which I suspect is a ghost.) The sequence of titles however is 'Manuel', 'Nouveau Manuel', 'Nouveau Manuel Complet' and then 'Nouveau Manuel Complet' published in two parts. This would accord with the introduction to Part II of the 1865 edition which mentions that the work is now being split into two parts. The First Part being entitled 'Pyrotechnie Militaire' now being the 5th edition, and the Second Part being entitled 'Pyrotechnie Civile' here being published separately for the first time. The sub-title being, 'La confection et la disposition des artifices de joie et des feux de couleur en usage dans les rejouissances publiques les fêtes particulaires au theatre et sur l'eau.' It is unfortunate that the 'grand nombre de livres français et étrangers relatifs au sujet que nous y traitons' that both the introduction and the index of the 1838 edition promised, was not actually included. The popularity of these works of Roret is such that a number of them have recently been reprinted by Léonce Laget, including the 1838 edition listed above. This has been extensively reviewed in *Pyrotechnica No.VIII*, June 1982, where Michael Swisher comments that, 'This work fairly reflects the state of pyrotechnic knowledge in 1838, not perhaps among the most advanced researchers, but among the average practitioners of the trade.'

References: Du Pont. CP (1838, 1981). Brock (1865)

VOGEL, Heinrich

Kurtzer Bericht der Artillerie Wissenschaft ...

V050.1
Zweyten Aus. David Gessner: Zurich 1739
EngTP+TP+136(-2)+9Pls. 17cms. NV — BL 8825.bbb.40

Kurtzer Unterricht in der Artillerie-Wissenschaft, zu Ernst-und Lust-feuerwerken. [*Short lessons in the science of artillery, for military and recreative fireworks.*]

V050.2
3. und vermehrte Aufl. Gessner: Zurich 1756
xxii+279+10FPl. 20cms. NV 0214126 BL —

The 2nd edition of 1739 does not contain any section on recreative fireworks, only 'Kunst und Ernst Feuerwerks'.

References: Du Pont

VOLLENKOMMENE Unterweisung See SYLVIUS, Frederick

W, J See WHITE, John

WANDROWETZ, A

Der Kunstfeuerwerker. Leichtfassliche, anleitung zum anfertigen einfacher aber schöner effektstücke. [*The fireworker. Easy instructions for making simple but beautiful fireworks.*]

W010.1
Hachmeister und Thal: Leipzig (1928)
TP+ii+36inc.26illus. 16cms NW — BL —

W010.2
Hachmeister und Thal: Leipzig 1939
NW — BL —

References: GV (1928). Pokorny's collection (1939). Barfod

WATKINS, Thomas Frederick, CACKETT, J.C & HALL, R.C

Chemical Warfare, Pyrotechnics and the Firework Industry.

W020.1
Pergamon Press: Oxford/London 1968
PC+½T+TP+vi+114(-2)2figs. NW — BL —

One of a series of monographs, prepared for senior students, on the chemical industry in the U.K. The third section on the Firework Industry by R.C. Hall occupies only 22 pages and gives a brief and simplistic survey of its history and general nature. The bibliography only contains six items.

References: Lancaster

WEBER, Carl

Die Lustfeuerwerkerei; oder, Vollstandige Anweisung zur Anfertigung aller Feuerwerkskorper ... Nebst practischer Anweisung zur Erzeugang des electrischen Lichtes, des chinesischen Feuerwerks, bengalischer Flammen, Lichter mit verschiedenen Farben & und Erklarung uber die verschiedenen Ingredienzien ... [*The making of fireworks; or, complete instructions for making firework items ... together with practical instructions for making electric lights, chinese fireworks, Bengal fires, lights in different colours and an explanation of the different ingredients.*]

W030.1
S. Mode: Berlin: c.1866
1Pl+TP+iv+148+ivads. NW — BL 8715.aaa.13

W030.2-11
Verm. Aufl. 6th 1875, 9th 1884, 11th 1892

References: Du Pont. Kayser. G&O (1884)

WEBSKY, Martin (1824-1886)

Grundliche und fasstliche Anweisung zur Verfertigung kleiner Lustfeuerwerke, zunachst für Dilettanten der Feuerwerkskunst. [*Basic and easy instructions for the manufacture of small fireworks, primarily for amateurs of the firework art.*]

W040.1
Landsh: Breslau 1831
2Pls. NW — BL —

Praktisches Handbuch der Lustfeuerwerkkunst, nach eigenen Ansichten und Erfahrungen bearbeitet von Martin Websky ... [*Practical handbook on the art of making fireworks, according to his own views and experience, edited by Martin Websky.*]

W040.2
(2nd ed.) August Schulz: Breslau 1834
viii+104+2FPl. 20cms. NW — BL 1044.h.20

Lustfeuerwerkerei für Dilettanten ...

W040.3
(3rd ed.) Ferdinand Hirt: Breslau 1839
vii+132+10Pls. 23cms. NW — BL —

Lustfeuerwerkkunst; oder, Leicht fassliche und bewahrte Anweisung. Zur Verfertigung von Lustfeuerwerken.

W040.4
4e umgear. Ausg. Ferdinand Hirt: Breslau 1842
288p. 24cms. NW 0138246 BL M.L.c.14

Meine neuesten Ansichten und Erfahrungen im Gebiete der Lustfeuerwerkkunst. Nachtrage zur vierten Ausg. der Lustfeuerwerkerei für Dilettanten.

W040.5
Ferdinand Hirt: Breslau 1844. (Bound with 1842 ed. in NUC)
½T+TP+xii+46p. NW 0138246 BL —

Martin Websky's Lustfeuerwerkerei insbesondere für Dilettanten und Freunde dieser Kunst. Nebst einem Doppel-Anhang: (I) Die Stoffe und Substanzen der Lustfeuerwerkerei ... von Ernst Schnell. (II) Auszug des Neuen und Beachtenswerthen aus F.M. Chertier's Lustfeuerwerkerei Nach dem Französischen mit Anmerkungen von Martin Webssky [*Martin Websky's making of recreative fireworks especially for amateurs and friends of this art. Together with double appendix; (I) The materials and substances of making fireworks ... Ernst Schell. (II) Extract of what is new and noteworthy in F.M. Chertier's The Making of Fireworks from the French with notes by Martin Websky.*]

W040.6
5e neuesten ... Ausg. Ferdinand Hirt: Breslau 1846
TP+TP+Eng+ii(Nachtrage I)+46+TP(Nachtrage II)+50+TP(Stoff und Substanzen) +(53-92)+ TP(F.M. Chertier)+(93-128). 21cms.
NW — BL —

Martin Websky's Lustfeuerwerkkunst, oder leicht fassliche, bewährte Anleitung zur Berfertigung von Lustfeuerwerken ... insbesondere für Dilettanten und Freunde der Lustfeuerwerkerei.

W040.7
6e. bearb. Ferdinand Hirt: Breslau 1858
TP+EngTP+320illus. 21cms. NW — BL —

Martin Websky's Lustfeuerwerkskunst ...

W040.8
7e. bearb. Ferdinand Hirt: Breslau und Leipzig 1873
viii+272(64illus.)+1Pl. NW — BL 7942.e.11

W040.9
Zweite Ausgabe der siebenten bear ... herausgegeben von Julius Hutstein und Prof. Dr. Martin Websky. F. Hirt: Leipzig 1878
Eng+TP+vi+272(64illus). 19cms. NW 0138244 BL —

W040.10
8 Auflage ... vollstandig umgearbeitet und erweitert von Tassilo Giesl v. Gieslingen. A. Hartleben: Wien 1891
vii+240illus+FTs. 21cms. NW 0138245 BL —

OTHER WORKS

Martin Websky's Schule der Lustfeuerwerkerei. Ein Vermachtniss an die Freunde und Dilettanten dieser Kunst ...

W040.11
Ferdinand Hirt: Breslau 1850
viii+255illus. 17cms. NW 0138249 BL 1379.b.19

W040.12
Attenburg: Breslau und Leipzig 1873
viii+272p. NW — BL 7942.e.11

Zur Lustefeuerwerkerei. Aus der Mappe seiner Erfahrungen mitgetheilt von Martin Websky.

W040.13
Ferdinand Hirt: Breslau 1854
46p. 17cms. NW — BL 8715.a.54/4

Websky was without doubt the most influential and prolific German writer of the 19th century on the subject. In many ways his work paralleled that of CHERTIER in France (whose work he frequently quoted and also translated) and several formulae used today are based on his writings. Details of all the above editions are to be found in the excellent Kayser. It will be seen that essentially there was one book with slightly differing titles that, together with a few additional parts, went through eight editions between 1831 and 1891. I have not however been able to discover if both father and son were called 'Martin' since the first edition of 1831 would have been publshed when one or other Martin was only 7 years old! Two other works were also published, the 'Schule' in 1850 and 1873 and the 'Mappe' in 1854.

References: Du Pont. Brock (1850, 1873). Kayser. CP (1846, 1858, 1878)

WEINGART, George Washington (1871-1948)

Dictionary of Pyrotechnics. A complete Manual covering the authors work and experiments from 1890 to 1930. [*(Cover) 'Dictionary and Manual of Pyrotechny'*.]

W050.1
(New Orleans 1930?)
TP+155s.s.+6Pls(Typescript). 27cms. NW 0161248 BL —

W050.2
(Reprint) Max P. Vander Horck: San Diego 1976?
TP+x+244. 28cms. NW — BL —

Dictionary and Manual of Pyrotechny, covering the authors work and experiments from 1890 to 1935. [*(Coloured cover) 'Dictionary and Manual of Fireworks - Weingart'*.]

W050.3
Copyright New Orleans 1937
TP+170(-2)+vi+4Col.Pls.85figs. 23cms. NW 0161247 BL 8716.f.29

Dictionary and Manual of Fireworks, by George W. Weingart, covering the author's work and experiments from 1890 to 1935 ...

W050.4
Bruce Humphries: Boston 1937
170p incl.4 col.Pls. 22cms. NW — BL —

Pyrotechny: a practical manual for Manufacturers of Fireworks, Signals, Flares and pyrotechnic Displays.

W050.5
Chemical Publishing Co.: New York 1939
170illus. col.Pls. 21.5cms. NW 0161253 BL —

Pyrotechnics, Civil and Military.

W050.6
Chemical Publishing Co.: New York 1943
220p. NW 0161251 BL —

Pyrotechnics.

W050.7
Second ed. rev. and enl. Chemical Publishing Co.: New York (1947)
TP+x+244illus.+4Col.Pls. 22cms. NW 0161252 BL —

W050.8
Ed. Christie: New Jersey 1966
TP+x+244. 22cms. NW — BL —

W050.9
(Reprint) Second ed. rev. and enl. Chemical Publishing Co.: New York [1968].
TP+x+244. 22cms. NW — BL —

Although generally considered to be the father of modern American firework practice nevertheless even by the time his first book appeared in 1930 some of his observations were already out-of-date. His experience of American manufacturing techniques was mainly derived from his work prior to the First World War and, as a result, almost totally ignored the 'Italian' school who now dominate the US scene. Some of his observations suggest that he had merely observed and made guesses about how things were done rather than having experienced them first-hand.

The 1968 reissue of Pyrotechnics lacks the 4 coloured plates that graced the original issue in 1937. The popularity of this work amongst American pyrotechnists however has meant that there are a number of pirated editions and photocopied versions in circulation. The NUC entry numbers NW0161249 and NW0161250 appear to duplicate NW0161248.

References: Du Pont (1939). Brock (1930, 1947). NYPL (1930, 1937, 1937, 1939, 1943). CP (1930, 1947, 1968)

WELMAN, Henry J

Fireworks and how to make them.

W060.1

New York c.1900

NW — BL —

References: Brock

WESTEN, Wynant von (? -1636)

(TP) Het eerste (-Derde) Deel von Mathematische Vermaeck-Lyckheden ... (TP Pt.III) Te samen ghevoeght uyt eene versamelinge van verscheydene ghenoeghlichke ende vermakelijcke Inventien van konstighe vyer-wercken. [*The first to third part of Récréations Mathematiques ... (Pt.III) Brought together from a collection of various pleasant and amusing inventions of artificial Fireworks.*]

W070.1
Jacobus van Biesen: Arnhem 1636
235p. NL — BL —

W070.2
(2nd ed.) Jacobus van Biesen: Arnhem 1641
235p. NL — BL —

W070.3
(3rd ed.) Jacobus van Biesen: Arnhem M.DC.XLIV. (1644)
257+xiii+61+vii+42p. NL 0303581 BL —

W070.4
(4th ed.) Jacob van Biesen: Arnhem 1662
½T+TP+xii+242+xiv+TP+55(-2)+v+TP+34(-1)+ii.
NL — BL 8530.aa.12

W070.5
(5th ed.) Jacob van Biesen: Arnhem 1672/71
229+55+v+34+ii.
NL 0303583 BL —

W070.6
(6th ed. rev.) Weduw van Theunis Iacobsz: Arnhem 1673
229+xi+53+v+34+ii. NL 0303585 BL —

Dutch translation of *Récréations Mathematiques* by ETTEN. Pt.III only has 2 of the original 12 woodcuts. NUC has apparent duplicate listings under NL0303608/9.

References: Haan 5302/3

WHITE, John (? -1671)

A rich cabinet with variety of inventions; unlock'd and opened, for the recreation of Ingenious Spirits at their vacant houres. Being Receits and Conceits of several natures ... As also variety of Recreative Fire-works both for land, air, and water. And fire-works of service for sea and shore ... Collected by J.W. a lover of artificial conclusions. [*Long-titles vary slightly*]

W080.1
W. Gilbertson: [London] 1651
90L+WC. 14cms. NW — BL E1295(2)

W080.2
2nd ed. with some additions. W. Gilbertson: London 1653
90L+WC. 14cms. NW 0252429 BL D

W080.3
3rd ed. with many additions. W. Gilbertson: London 1658
ii+EngTP+TP+x+191+WC. 16cms. NW 0252431 BL C.175.m.28

W080.4
4th ed. For William Whitwood: London 1668
ii+EngTP+TP+x+190+ii+WC. 16cms. NW 0252433 BL Cup408.c.53

W080.5
4th ed. W. Gilbertson: London 1668
190+WC. 16cms. NW — BL —

W080.6
5th ed. For William Whitwood: London 1677
EngTP+TP+x+190+42. NW — BL —

W080.7
5th ed. William Whitwood: London 1684
viii+128+WC. 14.5cms. NW 0252435 BL D

W080.8
5th ed. W. Gilbertson: London 1684
196+WC. NW — BL —

W080.9
6th ed. W. Whitwood: London 1689
xiv+128+WC. NW 0252437 BL 1O36.c.6

A Rich Cabinet of Modern Curiosities, containing, 1. Many Natural and Artificial Conclusions. 2. Variety of Fireworks, of great service for the Sea and Land ... With a Multitude of other curiosities, the whole Adorned with above 40 Curious Cuts.

W080.10
For G. Conyers ... :[London] c.1700
72p+WC. 15cms. NW 0252427 BL 1037.b.28

Hocus-Pocus: or, A Rich Cabinet of Legerdemain Curiosities. Natural and Artificial conclusions ... 13. To make variety of Fireworks ...

W080.11
Sold at the Ring in Little-Britain: [London] c.1715
Pl+TP+96(-4)+WC. 14cms. NW 0252420 BL C.175.d.11

W080.12
Sold at the Ring in Little-Britain: [London] c.1715
Pl+TP+72(-4)+WC. 15cms. NW — BL 1651/50

Art's Treasury of Rareties, and curious Inventions In Two Parts ... Part II To which ... is added, A variety of Artificial Fireworks, with Divers other Curiosities.

W080.13
Printed for Robert Smith ... : Glasgow 1761
90p+WC. 14.5cms. NW 0252419 BL —

John White's Bibliography provides a number of problems, not only because the great popularity of his works created a large number of editions, but most of them are very rare and their cataloguing confusing. (In the BL catalogue, due no doubt to the large number of authors with this name, he is subtitled 'A Lover of Artificial Conclusions', which is his own description of himself on some of the title-pages.) In addition the various titles are all very similar and there is a marked lack of dates associated with them.

There are three main title sequences which overlap chronologically. The first is *A Rich Cabinet* ... whose first edition was 1651; this was followed by *Art's Treasury* ... which ran from 1688 to about 1773; finally *Hocus-Pocus* of about 1710-15. The latter is a much abbreviated version of *A Rich Cabinet*. The only edition of *Art's Treasury* to contain a section on Fireworks is the Glasgow edition of 1761. Fireworks do not appear in the 6th edition of 1773.

Hodgkin says of his edition of *A Rich Cabinet*, 'This book is a curious and interesting one, so popular as to have gone through five editions before 1689 ... the book furnishes so good an illustration of what may almost be described as a scientific chap-book that I must give a very short notice of its contents ... The first portion deals with all manner of little domestic dodges and tricks, among which we find our old friend the telling of the time of day by a ring a thread and a tumbler. 'The School of Artificial Fireworks' is very full of details and their manufacture, and the cuts are quaint.'

Included in this part, all editions have a small section sub-titled 'Necessary and serviceable Fireworks both for Land and Sea execution' which describe Fire-pikes, How to invest ships, and other warlike matters. In the first edition of 1651 this consisted of 17 pages, but by the 6th edition this had been cut to only 9, it no doubt being considered inappropriate in such a book! All editions contain virtually the same 40 or so pages devoted to recreative fireworks which include the 27 'quaint' woodcuts referred to by Hodgkin, although one of the sections, on 'Combatants' and its woodcut, is dropped from the 6th edition. The third and subsequent editions contain an additional 8 pages (88-96) entitled 'Short but certain rules for the making of all sorts of fireworks for recreation as Rockets, Fisgigs, Runners on the Line, Serpents, Stars, Firewheels, Clubs, Jack in a Box etc ... By W.R.'. Much of this material seems to duplicate what has gone before.

The Firework section in *Hocus-Pocus* occupies pages 49-56, (of the 96 page edition), and includes 6 woodcuts. It deals only with primitive 'Fire-Pikes', and 'Bullets of Wild-Fire'. The woodcuts and the text are virtually the same as that used for *A Rich Cabinet*. This section appears rather out of place in a book which treats, in adjacent paragraphs, of 'To Drive Away Mice' and 'To Make Hair Grow Quickly'.

The two editions of *Hocus-Pocus* quoted above both have identically worded title-pages, but the '72 page' version has slightly more text to each page. The last page refers, for no apparent reason, to an accident, 'In the Year 1712, or thereabouts, near Durham'. This clearly dates it about 1714 which is later than Hall suggests. On the obverse of the title-page is an advertisement for 'G.C.' Books, so the publisher is obviously G. Conyers

References: Wing W1787 & 1789-93A. Stott 684-693 & 701. Hall 298.7 & 288-302. CP (1668). Arber (1677, 1684). NYPL (1677, 1689)

WHITEHORNE, Peter (c.1550-1563)

(I) The Art of Warre ... (containing) (II) Certain waies for the orderyng of Souldiers in batteiray ... and more ouer, howe to make Saltpeter, Gunpowder and diuers sortes of Fireworks or Wilde Fyre.

W090.1

John Kingston for Nicolas Englande: London (I) MDLX (II) MDLXII (1560/1562)
48L+vi (Leaves misnumbered). 19cms. NW — BL C54.bb.17

W090.2

VV. VVilliamson: London 1573
48L+vi. 19cms. NW 0262649 BL C143.cc.8

W090.3
T. East for I. Wight: London 1558
48L+vi. 19cms. NW 0262652 BL —

No mention is made of recreative fireworks in this book; it deals only with the military uses of fireworks and Gunpowder. 'Whitehorne translates, without acknowledgement, the greater part of Book X of (Biringuccio's) *Pirotechnia*' (Smith & Gnudi).

WHOLE

The Whole Art of Legerdemain; or, Hocus Pocus in Perfection ... including The Newest Discoveries. To Which is Added ... The Art of Making Fire-Works.

W100.1
Printed and Sold by J. Bailey: [London] c.1810
ColPl+iii+28(-4). 19cms. NW 0272542 BL —

References: Stott 85. Hall 308

The Whole secret laid open; or, The Complete Art of making the chemical fulminating objects such as the lace, or girt of security, fulminating letters, balls, bombs, garters, cards, spiders, segars, chairs, drawers, boots, shoes &c. &c.

W110.1
W. Glindon: London 1814
34p. 17cms. NW — BL —

W110.2
J. Johnson for the author: London 1818
NW — BL —

The title of this work is so similar to those by GROTZ, that they are possibly all by the same author.

References: Du Pont (1814). NYPL 1818

WINTHER

De pulvere pyrio.

W120.1
Hafnia (Copenhagen) 1698
NW — BL —

Brock lists this under 'Winter'. It deals only with the production of gunpowder and it's constituent chemicals.

References: RDML. Brock

WOTHE, J

Der Kleine Feuerwerker.

W130.2
2e. aufl. Siwenna: Kattowitz 1885.
21p NW — BL —

References: G&O

YOKOSHIMA, Koichi (1908- ?)

Otsukado no hanabimatsuri. [*The Firework Festivals of Otsukado.*]

Y010.1
Ron Shobo: 1978
149p. 18cms. NY — BL —

Fireworks and festivals in religion and folklore.

References: NUC SC 1980

YOUNG

The Young Man's Book of Amusement. Containing the Most Interesting and Instructive Experiments in various Branches of Science. To which is Added all the Popular Tricks and Changes in Cards, and the Art of Making Fire Works.

Y020.1
Printed and Published by William Milner: Halifax MDCCCXXXIX (1839)
FPl+TP+384p. 13cms. NY 0032122 BL 7922.a.8

Stott records issues of 1840, 1844, 1846 & 1850. (He erroneously give MDCCCIX for the 1839 edition). Hall adds 1848 & 1859. NUC lists an undated version published by Milner in London as well as issues of 1846, 1850 & 1854. The Findlay sale at Sotheby in December 1982 contained a couple of copies dated 1852. *The Art of Making Fireworks* occupies pages 355-362 and deals extremely briefly with touch-paper, scrole Wheels, fires, sparks, squibs & crackers.

References: Stott 749-52 & 1216. Hall 320-23

Illustrations

246 Pratica Manuale di Artigl.

mancare in questa mia Prattica di Artiglieria in cosa alcuna appartenente à essa, noterai le infrascritte differenze de' fuochi, ancor che molte altre te ne poteua scriuere. Et prima, cominciando da i raggi, il cui uso per esser tanto commune, che insino à i putti gli sanno fare, & ancora fargli volare per vna corda, & andare innanzi, & tornar'indietro per essa, non farò particolari ragionamenti, per non fastidir troppo i Lettori. Ma non starò per questo di rapresentarti la sua figura, nella qual si dimostra in che modo si accommodano li Raggi da potergli far'andare di vna Torre à vn'altra per il lungo di vna corda, & il medesimo dirò delle piule, ouero diciamo tronadori di carta reale, fatte à molte pieghe, le quali ogn'vno prattico le sà fare. Dopò queste seguitano le Ruote, ouero Girandole, si come nella seguente figura vedrai, le quali etiandio sono molto vsate nelle Feste: Però il modo di farle, per esser troppo prolisso, non mi basta l'animo di esplicarlo. Dopò le Girandole ci sono altri diuersi modi da rallegrare, & trattenere la gente nel tempo delle allegrezze, & feste come saria à dire, quando tù con quegli istessi cannoni, che vuoi fare la salua, volessi fare spargersi in Cielo, & poi insino à terra vna spessa pioggia di scintille di fuoco, il che in vero fa vn vedere bellissimo, & gratioso, il che farai in questo modo: Piglia vna gran quantità di resegature grosse di Pino, ò di Lauro, ò di Pobbia, non potendo hauere d'altro, & mancando di queste, delle raspadure grosse del Souero ti puoi seruire, & quelle farai bollire per vn buon pezzo con acqua del Salnitro, & come haueranno bollito, caua la

Fig. 83.

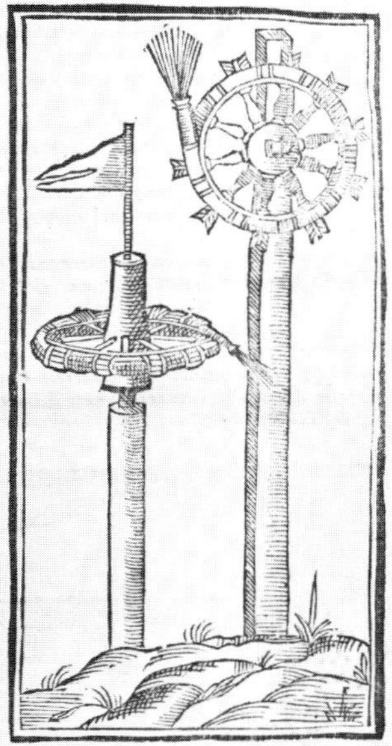

& modo di Fuochi artificiali. 247

la caldara dal fuoco, & fa scorrere quell'acqua in altro vaso, & caua fuora le resegature, & mettile stese sopra di vna tauola grande, & essendo ancora loro bagnate, & calde, poluerizale con poluere di Solfo trita, & setacciata sottilmente, & valle mescedando bene insieme, & poi lasciale seccare, & gouernale fin' al tempo, che te ne vuoi seruire; & volendo tù caricare li pezzi, dopò che tù hauerai data la carica ordinaria, prima di metter bottoni de sfilazzi, nè alcuna altra cosa, mettiui dentro due, ò trè cucchiare di quella resegatura, & calca benissimo, che vada incorporata appresso alla poluere, & allhora tù metterai il bottone, ò foraggio molto ben calcato. Fatto questo, dà fuoco al cannone al tempo debito, & vedrai effetto marauiglioso; percioche tutte quelle resegature si accendono dal fuoco, & sparse per l'aere, occupano vna gran parte d'esso per tutto intorno, & à modo di vna pioggia spessa di fuoco cascano, insin'à terra ardendo, & non per questo resta di far' la salua l'artiglieria; anzi fa più forte il tuono; però nota, accioche meglio si possa sentir' il tuono, conuiene, che tali pezzi stiano in luoghi alti, & eminenti. Accostumasi ancora in simil feste che si fanno delle granate, le quali la notte faranno vn bel vedere, &

Fig. 84

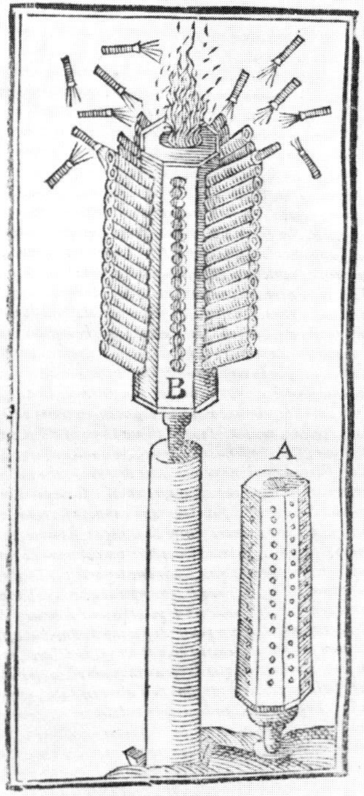

Plate 1

Collado, Luys. *Platica Manval de Artilleria.*

The first edition of 1586 did not include these wood-cuts of primitive wheels and a 'girandole' although they are described. They first appeared in the 1592 edition. These reproductions are from the 1641 edition, but are identical despite the 50 year gap!

PYROTECHNIA

HOC EST,

DE IGNIBVS FESTIVIS,

IOCOSIS, ARTIFICIALIBVS ET SERIIS, VARIISQVE EORVM ſtructuris.

LIBRI DVO.

EX SCRIPTORIBVS LATINIS,

Italis, & Germanis collecti, & in methodum ſuccinctam redacti.

In Gratiam eorum, qui hiſce artibus iucundi ſpectaculi cauſſa delectantur.

Auctore

ADRIANO ROMANO MATHEMATICO ac Profeſſore Herbipolenſi.

M. DC. XI.

PROSTAT
In Officina Paltheniana.

Plate 2
Title-page of Romanus, Adrianus *Pyrotechnia.* (1611)

Plate 3
Appier, Jean. *Recveil de plusieurs Machines Militaires.* (1620)

Plate depicting a typical 16th and 17th century 'naumachy' or sea-battle with fireworks.

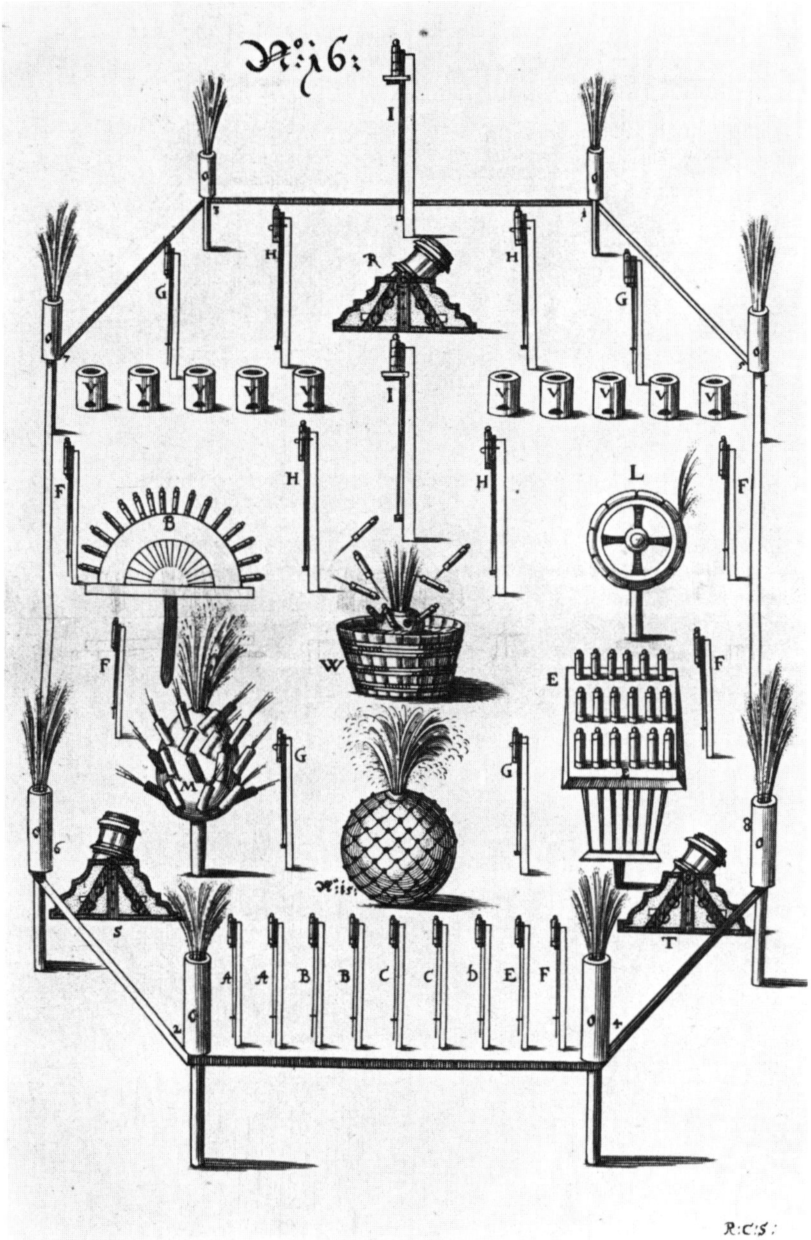

Plate 4
Furttenbach, Leonhard. *Halinitro-Pyrobolia.* (1627)

A layout of a typical firework display showing the disposition of the various items, compare this with the very similar plate (Plate 7) from Babington's *Pyrotechnia* published eight years later. It is interesting to note how similar in layout firework displays are today and how easily recognisable are the component parts.

RECREATIONS
MATHEMATIQVES.

Composées,

De plusieurs Problemes, plaisans & facetieux, d'Arithmetique, Geometrie, Astrologie, Optique, Perspectiue, Mechanique, Chymie, & d'autres rares & curieux Secrets : Plusieurs desquels n'ont iamais esté Imprimez.

PREMIERE ET SECONDE PARTIE.

La troisiesme Partie, contient vn Recueil de plusieurs gentilles & recreatiues inuentions de feux d'artifice : La maniere de faire toutes sortes de fuzées, simples & composées : Le tout representé par figures.

A ROVEN,
Chez CHARLES OSMONT, ruë aux Iuifs, prés le Palais.

M. DC. XXVIII.

Des fuzées qui sont portées par des cordes.

CHAP. VI.

IL y a diuerses façons de fuzées, qu'on fait voler sur des cordes, & ornées de plusieurs figures. Il y en a aussi de simples & de composées. Les simples sont emplies de leur composition,

Plates 5a & 5b
Etten, Henri van. *Recreations Mathematiqves.* (1628)

The title-page and woodcut from the first edition of this series to contain the section on fireworks. Virtually every work of a similar nature since that date has included chapters on fireworks. The woodcut depicts a flying dragon or line rocket, an item that has been popular in displays for over 400 years!

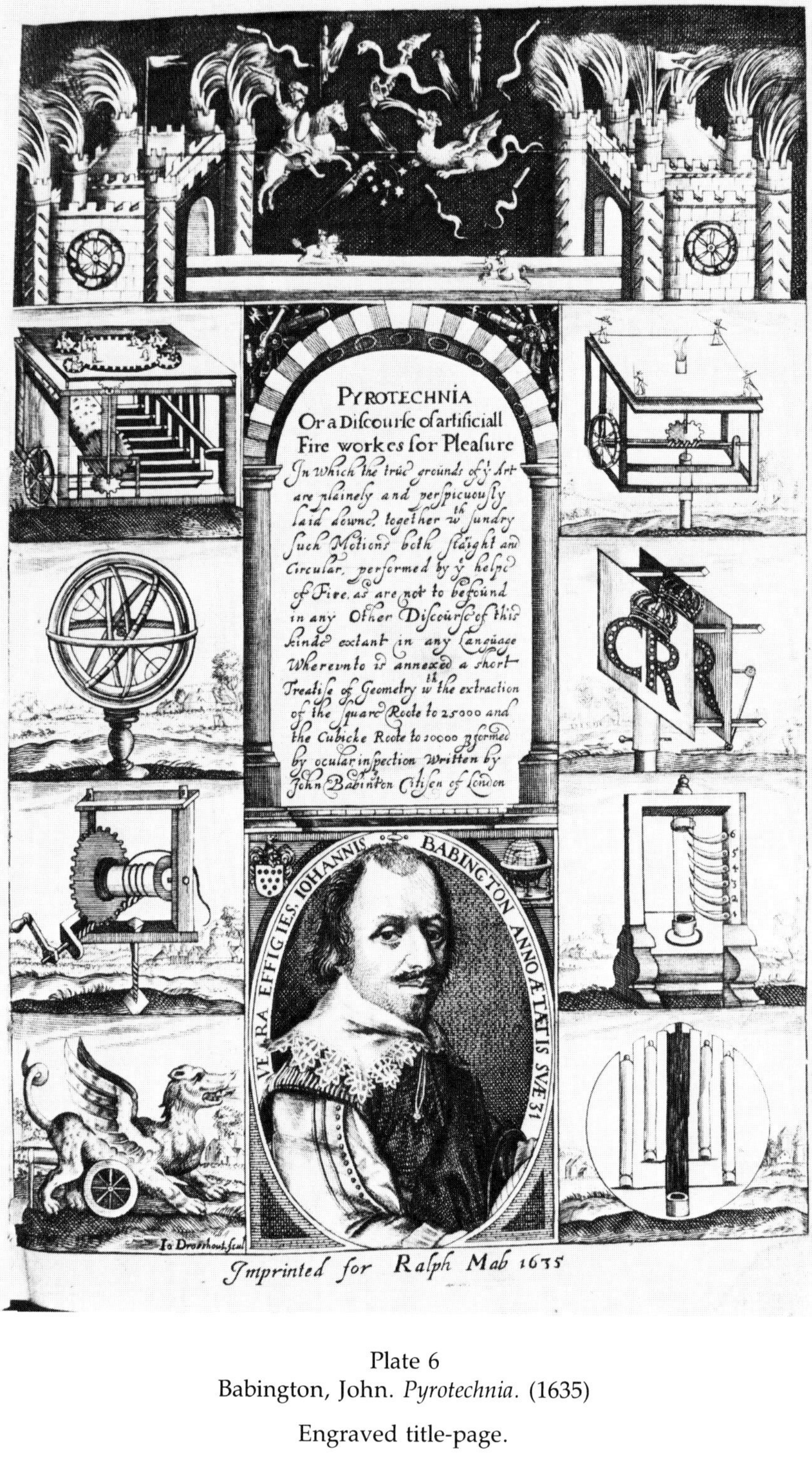

Plate 6
Babington, John. *Pyrotechnia.* (1635)

Engraved title-page.

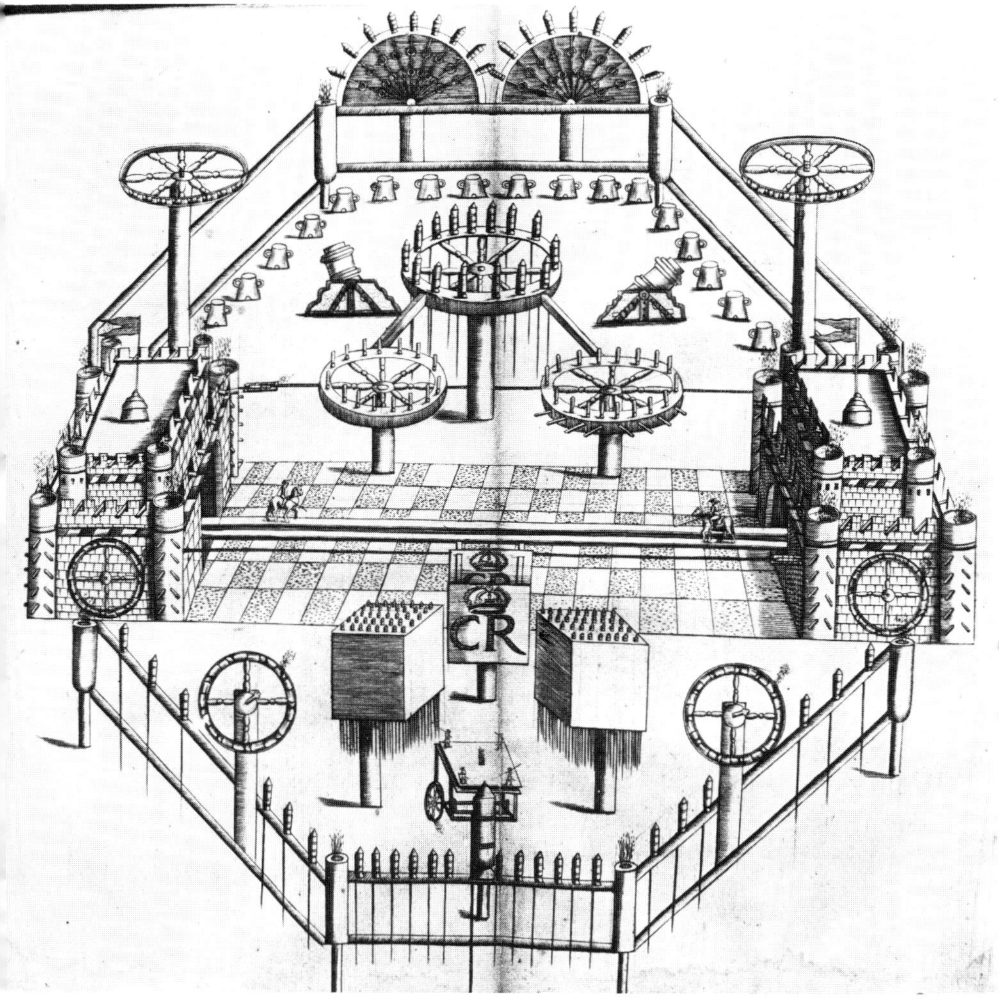

Plate 7
Babington, John. *Pyrotechnia.* (1635)

Layout of a typical display of the period. Compare with Plate 4.

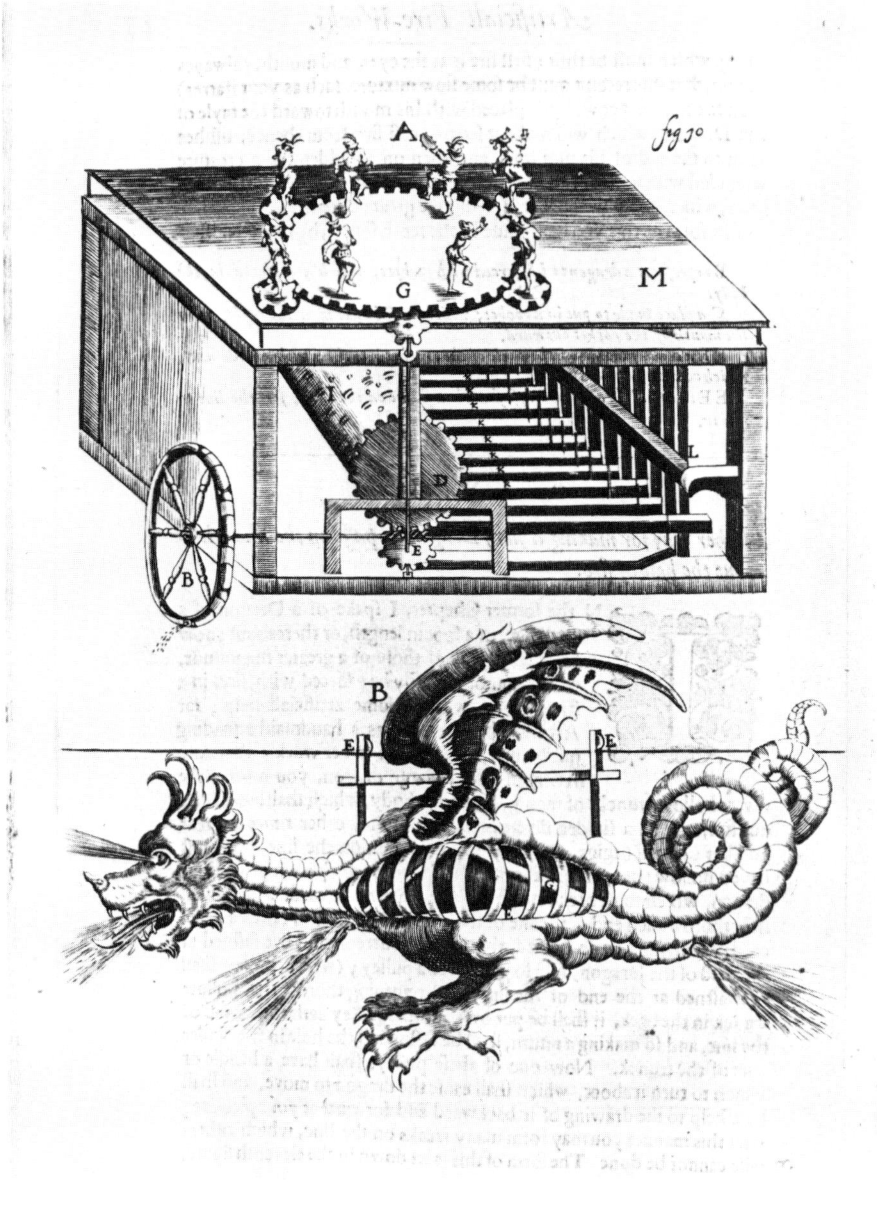

Plate 8

Babington, John. *Pyrotechnia.* (1635)

Plate depicting the use of rocket power to drive automata and a particularly fearsome dragon attached to a line rocket.

Plate 9
Simienowicz, Casimir. *Artis Magna Artilleriae.* (1650)

Engraved title-page.

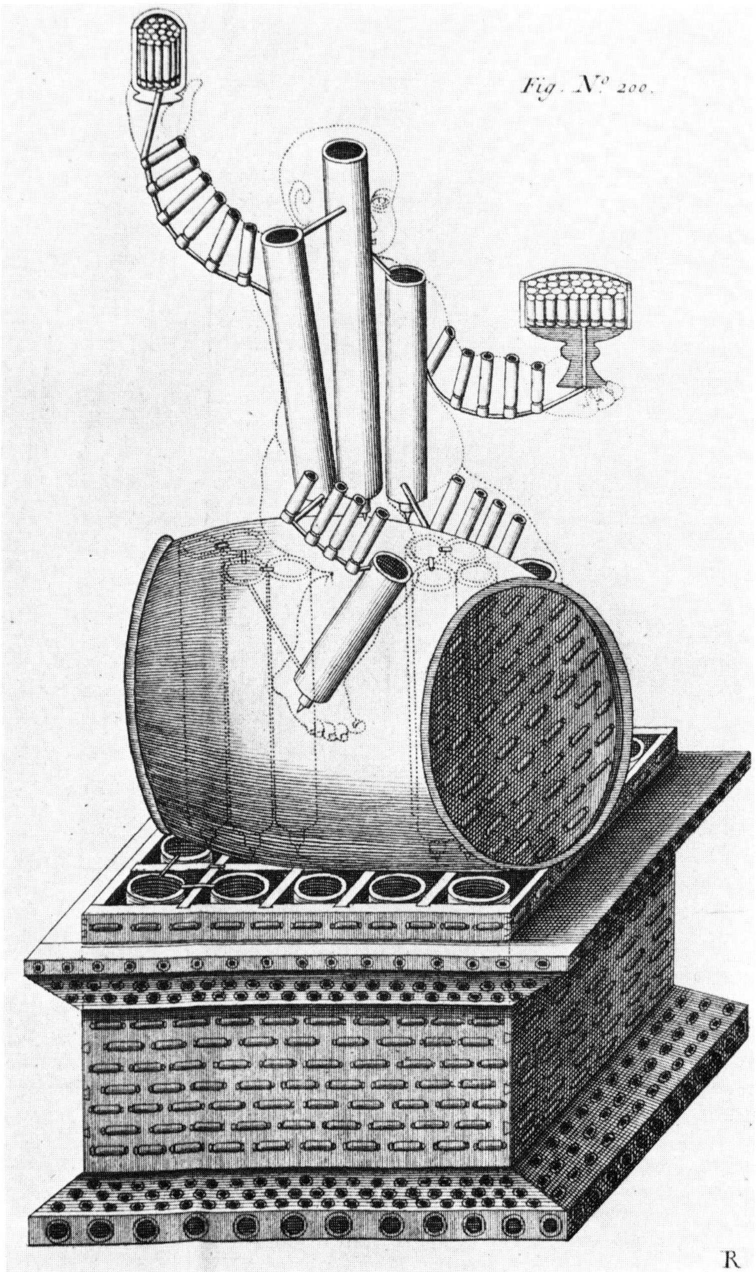

Plates 10 & 11
Simienowicz, Casimir. *Artis Magna Artilleriae.* (1650)

These two Plates clearly show how the firework machines and artifices, so common in the 17th, 18th and early 19th centuries, were constructed so as to hide the fireworks and surprise the audiences. The cost and elaboration of these items can no longer be afforded.

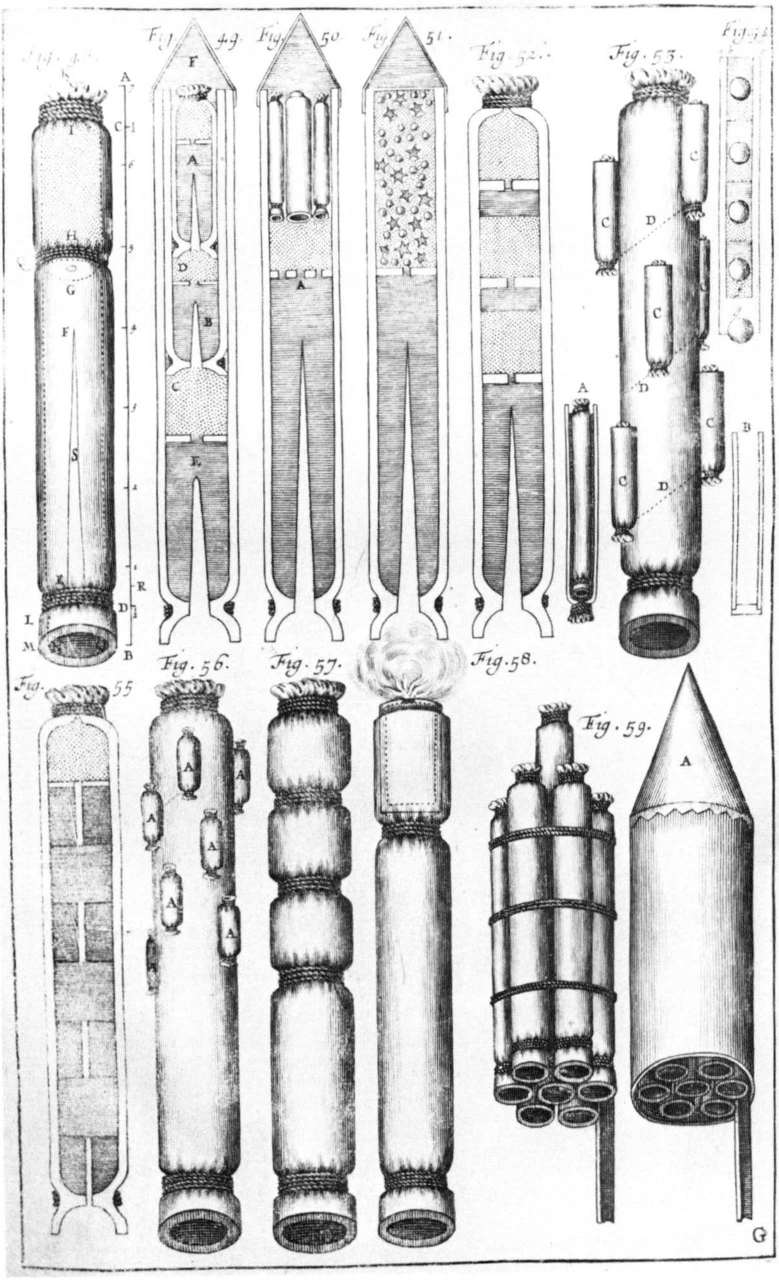

Plate 12

Sylvius, Frederick. *Unterischiedene neue Arten von Kunstlichen.* (1657)

This plate depicts the many ingenious types of rocket that were being made at the time, including three stage rockets (second from the left, top row). Such elaboration and complexity is rare today.

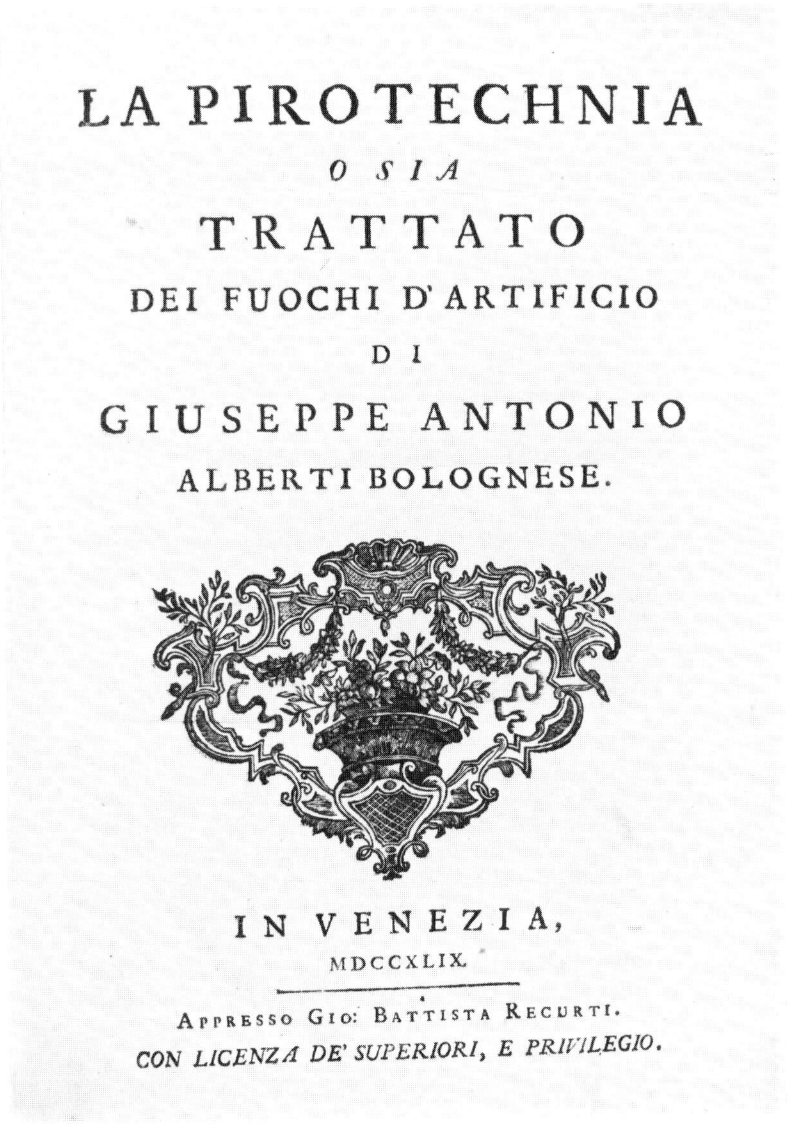
LA PIROTECHNIA

O SIA

TRATTATO

DEI FUOCHI D'ARTIFICIO

DI

GIUSEPPE ANTONIO

ALBERTI BOLOGNESE.

IN VENEZIA,

MDCCXLIX.

APPRESSO GIO: BATTISTA RECURTI.

CON LICENZA DE' SUPERIORI, E PRIVILEGIO.

Plate 13

Alberti, Giuseppe. *La Pirotechnia.* (1749)

The title-page of the first work in Italian devoted solely to recreative fireworks.

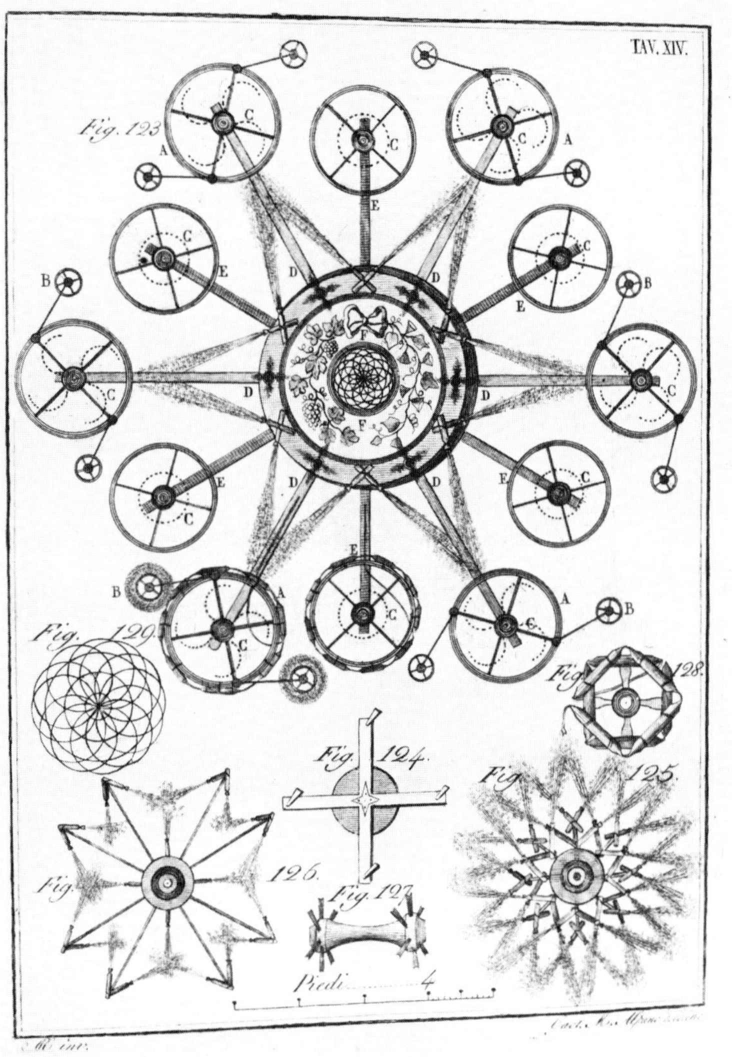

Plate 14
Ossorio, Marcello. *Istituzioni di Pirotechnia.* (1819)

Very elaborate set-pieces which have always been a speciality of the Italian school and can still be seen in some parts of the Mediterranean, especially Malta, even today.

A

SYSTEM OF PYROTECHNY,

COMPREHENDING THE THEORY AND PRACTICE, WITH THE APPLICATION OF CHEMISTRY;

DESIGNED FOR EXHIBITION AND FOR WAR.

IN FOUR PARTS:

CONTAINING AN ACCOUNT OF THE SUBSTANCES USED IN FIRE-WORKS; THE INSTRUMENTS, UTENSILS, AND MANIPULATIONS; FIRE-WORKS FOR EXHIBITION; AND MILITARY PYROTECHNY.

ADAPTED TO THE

MILITARY AND NAVAL OFFICER, THE MAN OF SCIENCE, AND ARTIFICER.

BY JAMES CUTBUSH, A. S. U. S. A.

LATE ACTING PROFESSOR OF CHEMISTRY AND MINERALOGY, IN THE UNITED STATES MILITARY ACADEMY—MEMBER OF THE AMERICAN PHILOSOPHICAL SOCIETY—CORRESPONDING MEMBER OF THE COLUMBIAN INSTITUTE—MEMBER OF THE LINNÆAN AND AGRICULTURAL SOCIETIES OF PHILADELPHIA—LATE PRESIDENT OF THE COLUMBIAN CHEMICAL SOCIETY, AND VICE-PRESIDENT OF THE SOCIETY FOR THE PROMOTION OF A RATIONAL SYSTEM OF EDUCATION, &c. &c. &c.

PHILADELPHIA:
PUBLISHED BY CLARA F. CUTBUSH.
1825.

Plate 15
Cutbush, James. *A System of Pyrotechny.* (1825)

The first American work on fireworks.

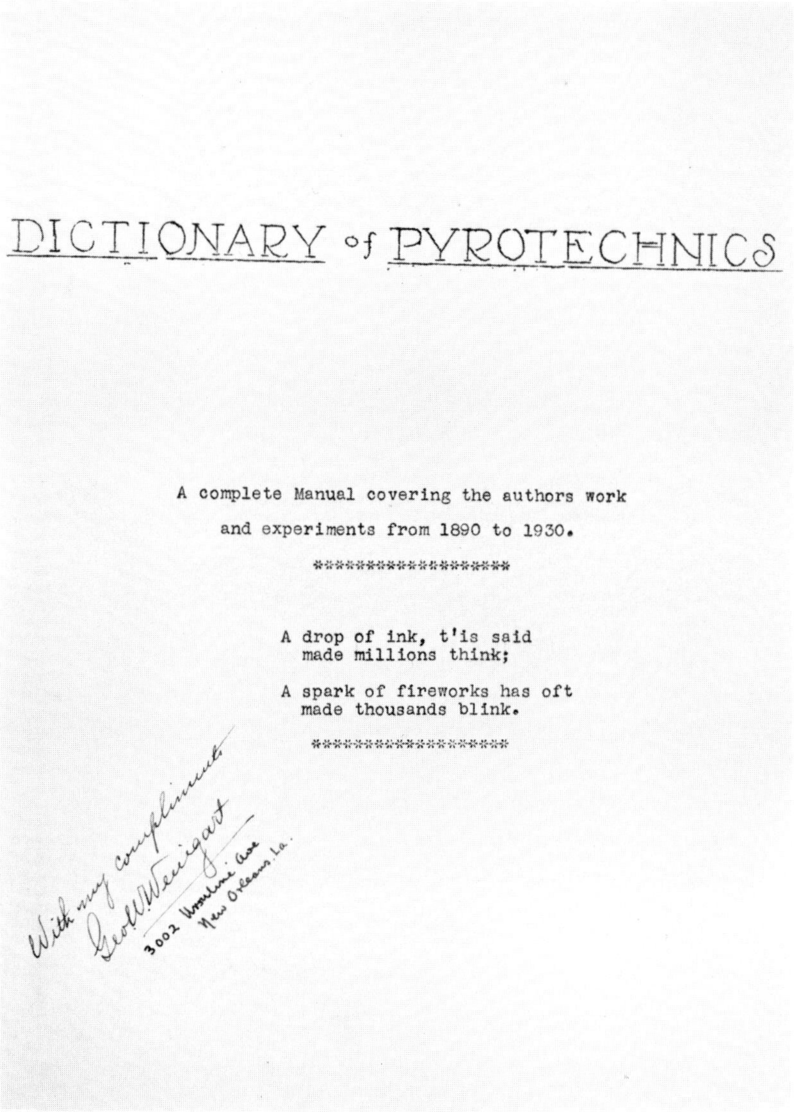
DICTIONARY of PYROTECHNICS

A complete Manual covering the authors work
and experiments from 1890 to 1930.

****

A drop of ink, t'is said
made millions think;

A spark of fireworks has oft
made thousands blink.

****

With my compliments
Geo W. Weingart
3002 Ursuline Ave
New Orleans, La.

Plate 16

Weingart, George. *Dictionary of Pyrotechnics.* (1930)

Autographed title-page from the first typescript work of the 'father' of modern American pyros.

The Juvenile book for Making Fireworks - - - 12mo. (c 1830)

Frontis (folding) by CRUIKSHANK. DAVID MILES 1988 £350

OPPENHEIMER 1383 . COHN 317.